Everyday Mathematics®

The University of Chicago School Mathematics Project

Student Math Journal
Volume 2

Grade

 Education

Chicago, IL • Columbus, OH • New York, NY

The University of Chicago School Mathematics Project (UCSMP)

Max Bell, Director, UCSMP Elementary Materials Component; Director, *Everyday Mathematics* First Edition; James McBride, Director, *Everyday Mathematics* Second Edition; Andy Isaacs, Director, *Everyday Mathematics* Third Edition; Amy Dillard, Associate Director, *Everyday Mathematics* Third Edition; Rachel Malpass McCall, Associate Director, *Everyday Mathematics* Common Core State Standards Edition

Authors

Max Bell, John Bretzlauf, Amy Dillard, Robert Hartfield, Andy Isaacs, Rebecca W. Maxcy†, James McBride, Kathleen Pitvorec, Peter Saecker, Robert Balfanz*, William Carroll*, Sheila Sconiers*

*First Edition only   †Common Core State Standards Edition only

Technical Art

Diana Barrie

Photo Credits

Cover (l)Tony Hamblin/Frank Lane Picture Agency/CORBIS, (r)Gregory Adams/Lonely Planet Images/Getty Images, (bkgd)John W. Banagan/Iconica/Getty Images; **Back Cover Spine** Gregory Adams/Lonely Planet Images/Getty Images; **iii iv v vi** The McGraw-Hill Companies; **vii** PhotoDisc/Getty Images; **189** The McGraw-Hill Companies.

Third Edition Teachers in Residence

Rebecca W. Maxcy, Carla L. LaRochelle

UCSMP Editorial

Laurie K. Thrasher, Kathryn M. Rich

Contributors

Martha Ayala, Virginia J. Bates, Randee Blair, Donna R. Clay, Vanessa Day, Jean Faszholz, James Flanders, Patti Haney, Margaret Phillips Holm, Nancy Kay Hubert, Sybil Johnson, Judith Kiehm, Carla LaRochelle, Deborah Arron Leslie, Laura Ann Luczak, Mary O'Boyle, William D. Pattison, Beverly Pilchman, Denise Porter, Judith Ann Robb, Mary Seymour, Laura A. Sunseri

 This material is based upon work supported by the National Science Foundation under Grant No. ESI-9252984. Any opinions, findings, conclusions, or recommendations expressed in this material are those of the authors and do not necessarily reflect the views of the National Science Foundation.

---

everyday**math**.com

STEM McGraw-Hill is committed to providing instructional materials in Science, Technology, Engineering, and Mathematics (STEM) that give all students a solid foundation, one that prepares them for college and careers in the 21st century.

Send all inquiries to:
McGraw-Hill Education
STEM Learning Solutions Center
P.O. Box 812960
Chicago, IL 60681

ISBN: 978-0-07-657642-5
MHID: 0-07-657642-6

Printed in the United States of America.

4 5 6 7 8 9 QDB 17 16 15 14 13 12

The McGraw·Hill Companies

Contents

UNIT 7 Fractions and Their Uses; Chance and Probability

UNIT 8 Perimeter and Area

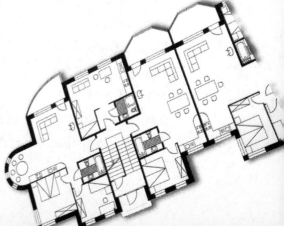

UNIT 9 Fractions, Decimals, and Percents

UNIT 10 Reflections and Symmetry

UNIT 11 3-D Shapes, Weight, Volume, and Capacity

UNIT 12 Rates

Projects

Activity Sheets

Date _____   Time _____

Fraction Review

Divide each shape into equal parts. Color a
fraction of the parts. Write the name of the
"whole" in the **"whole" box.**

1.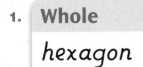
Whole
hexagon

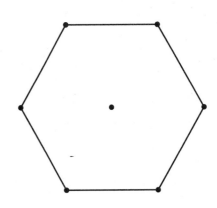

Divide the hexagon into 2 equal parts.
Color $\frac{1}{2}$ of the hexagon.

2. | Whole |
| --- |
| |

Divide the rhombus into 2 equal parts.
Color $\frac{0}{2}$ of the rhombus.

3. | Whole |
| --- |
| |

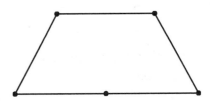

Divide the trapezoid into 3 equal parts.
Color $\frac{2}{3}$ of the trapezoid.

4. | Whole |
| --- |
| |

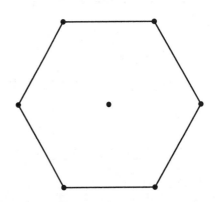

Divide the hexagon into 3 equal parts.
Color $\frac{1}{3}$ of the hexagon.

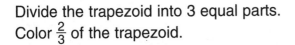

185

LESSON 7·1   **Fraction Review** *continued*

5. **Whole**

Divide the hexagon into 6 equal parts.
Color $\frac{5}{6}$ of the hexagon.

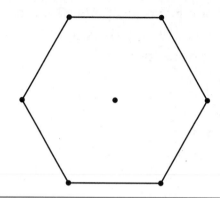

6. **Whole**

Divide each hexagon into thirds.
Color $1\frac{2}{3}$ hexagons.

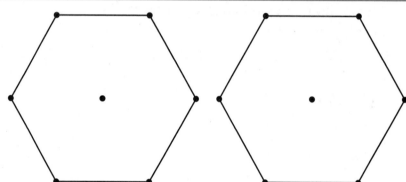

7. **Whole**

Divide each rhombus into 2 equal parts.
Color $2\frac{1}{2}$ rhombuses.

8. Grace was asked to color $\frac{2}{3}$ of a hexagon.
This is what she did. What is wrong?

_____

_____

_____

_____

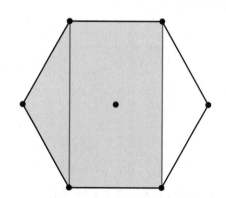

LESSON 7·1   Fraction Review *continued*

Fill in the missing fractions and mixed numbers on the number lines.

9.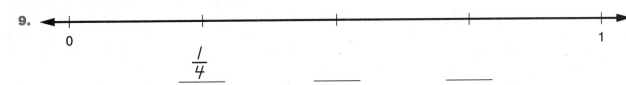

0                          $\frac{1}{4}$ _____   _____   1

10.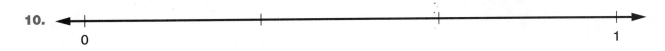

0   _____   _____   1

11.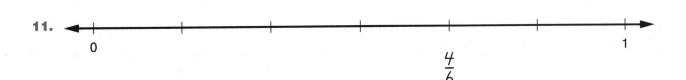

0   _____   _____   _____ $\frac{4}{6}$ _____   1

12.

0   _____   _____   _____   _____   _____   _____   1

13.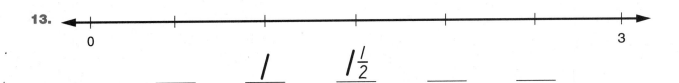

0   _____ 1 $1\frac{1}{2}$ _____   _____   3

14.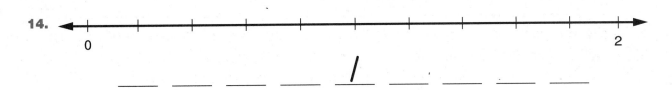

0   _____   _____   _____ 1 _____   _____   _____   2

Try This

15. Enter the fractions above on your calculator. Record the keystrokes you used to enter $\frac{2}{4}$ and $1\frac{1}{5}$.

_____

_____

_____

_____

187

LESSON
7·1

Math Boxes

---

1. What fraction of the clock face is shaded?

_____

SRB
56

---

2. ∠POL is an _____ (acute or obtuse) angle.

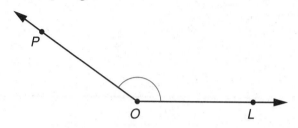

The measure of ∠POL is

_____ °.

SRB
93 142
143

---

3. Multiply. Use a paper-and-pencil algorithm.

_____ = 94 * 34

SRB
18 19

---

4. The five largest birds that are able to fly have the following weights: 16.3, 16.8, 20.9, 15.8, and 15.8 kilograms.

a. What is the median weight? _____ kg

b. What is the mode? _____ kg

c. What is the range? _____ kg

d. What is the mean? _____ kg

SRB
73 75

---

5. **a.** What city in Region 1 is located near 30°N latitude and 31°E longitude?

b. In which country is the city located?

c. On which continent is the city located?

_____

SRB
282 283

---

6. **a.** Measure and record the length of each side of the rectangle.

_____ in.

_____ in.   _____ in.

_____ in.

b. What is the total distance around the rectangle called? Circle one.

perimeter   area

SRB
131

LESSON 7·2   "Fraction-of" Problems

1.
Whole

16 nickels

a. Circle $\frac{3}{4}$ of the nickels.

b. How much money is that?

$_____._____

2.
Whole

a. Fill in the "whole" box.

b. Circle $\frac{5}{6}$ of the dimes. How much money is that?

$_____._____

3.
Whole

a. Fill in the "whole" box.

b. Circle $\frac{3}{5}$ of the quarters. How much money is that?

$_____._____

LESSON
7·2   **"Fraction-of" Problems** *continued*

Solve.

4. a. $\frac{1}{3}$ of 12 = _____   b. $\frac{2}{3}$ of 12 = _____   c. $\frac{5}{3}$ of 12 = _____

5. a. $\frac{1}{5}$ of 15 = _____   b. $\frac{3}{5}$ of 15 = _____   c. $\frac{7}{5}$ of 15 = _____

6. a. $\frac{1}{4}$ of 36 = _____   b. $\frac{3}{4}$ of 36 = _____   c. $\frac{6}{4}$ of 36 = _____

7. a. $\frac{1}{8}$ of 32 = _____   b. $\frac{5}{8}$ of 32 = _____   c. $\frac{9}{8}$ of 32 = _____

8. a. $\frac{1}{6}$ of 24 = _____   b. $\frac{4}{6}$ of 24 = _____   c. $\frac{13}{6}$ of 24 = _____

9. $\frac{2}{4}$ of 14 = _____

10. $\frac{2}{4}$ of 22 = _____

11. What is $\frac{1}{2}$ of 25? _____ Explain.

_____

_____

12. Michael had 20 baseball cards. He gave $\frac{1}{5}$ of them to his friend Alana,
    and $\frac{2}{5}$ to his brother Dean.

    a. How many baseball cards did he give to Alana? _____ cards

    b. How many did he give to Dean?   _____ cards

    c. How many did he keep for himself?   _____ cards

Try This

13. Maurice spent $\frac{1}{2}$ of his money on lunch. He has $2.50 left.

    How much money did he start with? _____

14. Erika spent $\frac{3}{4}$ of her money on lunch. She has $2.00 left.

    How much money did she start with? _____

LESSON 7·2

Math Boxes

---

1. What fraction of the clock face is shaded? Fill in the circle next to the best answer.

Ⓐ $\frac{1}{3}$

Ⓑ $\frac{6}{12}$

Ⓒ $\frac{1}{4}$

Ⓓ $\frac{2}{1}$

SRB 56

---

2. Draw angle *ABC* that measures 65°.

∠*ABC* is an _____ (acute or obtuse) angle.

SRB 93 142 143

---

3. Mary has 27 pictures. She gives $\frac{1}{3}$ of them to her sister Barb and $\frac{2}{3}$ to her cousin Sara.

a. How many pictures does Barb get?

_____ pictures

b. How many pictures does Sara get?

_____ pictures

c. How many pictures does Mary keep?

_____ pictures

SRB 59

---

4. Divide. Use a paper-and-pencil algorithm.

962 / 12 = _____

SRB 22 23 179

---

5. There are 29 students in Ms. Wright's class. Each collected 50 bottle caps. How many bottle caps did the students collect in all?

_____ bottle caps

---

6. Find the area of the figure.

☐ = 1 square centimeter

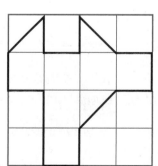

Area = _____ square cm

SRB 133

LESSON 7·3 Playing Card Probabilities

A deck of regular playing cards is placed in a bag. You shake the bag and, without looking, pick one card.

1. How many possible outcomes are there?
   (*Hint:* How many cards are in the bag?) _____ possible outcomes

2. Are the outcomes equally likely?
   (*Hint:* Does each card have an equal chance of being chosen?) _____

3. Find the probability of each event. Probability of an event $= \dfrac{\text{number of favorable outcomes}}{\text{number of possible outcomes}}$

Event	Favorable Outcomes	Possible Outcomes	Probability
Pick a red card	26	52	$\dfrac{26}{52}$
Pick a club		52	$\dfrac{}{52}$
Pick a non-face card		52	
Pick a diamond face card		52	
Pick a card that is *not* a diamond face card		52	
Pick the ace of clubs		52	
Pick a red *or* a black card		52	
Pick the 23 of hearts		52	

4. Circle the word or phrase that best describes the probability of picking a 5 from a bag of 52 regular playing cards without looking.

   impossible   very unlikely   even chance   likely

   Explain why you chose your answer. _____

Math Boxes

---

1. What fraction of the clock face is shaded?

   _____

   SRB
   56

2. ∠MRS is an _____ (acute or obtuse) angle.

   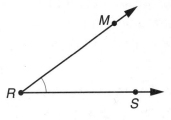

   The measure of ∠MRS is
   °
   _____ .

   SRB
   93 142
   143

---

3. Multiply. Use a paper-and-pencil algorithm.

   _____ = 19 * 473

   SRB
   18 19

4. Cleo's friends ran the 50-yard dash in the following times:

   7.9, 12.1, 8.5, 11.7, 8.3, 11.7, and 9.8 seconds.

   What is the mean time? Fill in the circle next to the best answer.

   Ⓐ 11.7 seconds

   Ⓑ 9.8 seconds

   Ⓒ 10 seconds

   Ⓓ 12.1 seconds

   SRB
   73–75

---

5. a. What city in Region 2 is located near 60°N latitude and 10°E longitude?

   _____

   b. In which country is the city located?

   _____

   c. On which continent is the city located?

   _____

   SRB
   284 285

6. Measure the length and width of your journal to the nearest half-inch. Find its perimeter.

   a. Length = _____ inches

   b. Width = _____ inches

   c. Perimeter = _____ inches

   SRB
   131

LESSON 7·4 Pattern-Block Fractions

Use *Math Masters,* page 212. For Problems 1–6, Shape A is the whole.

Whole

Shape A: small hexagon

1. Cover Shape A with trapezoid blocks. What fraction of the shape is covered by 1 trapezoid? _____

2. Cover Shape A with rhombuses. What fraction of the shape is covered by

   1 rhombus? _____

   2 rhombuses? _____

3. Cover Shape A with triangles. What fraction of the shape is covered by

   1 triangle? _____

   3 triangles? _____

   5 triangles? _____

4. Cover Shape A with 1 trapezoid and 3 triangles. With a straightedge, draw how your shapes look on the hexagon at the right. Label each part with a fraction.

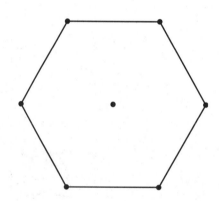

5. Cover Shape A with 2 rhombuses and 2 triangles. Draw the result on the hexagon below. Label each part with a fraction.

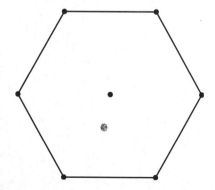

6. Cover Shape A with 1 trapezoid, 1 rhombus, and 1 triangle. Draw the result on the hexagon below. Label each part with a fraction.

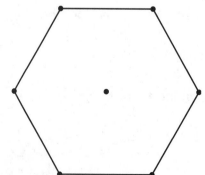

LESSON 7·4 **Pattern-Block Fractions** *continued*

Use *Math Masters,* page 212. For Problems 7–12, Shape B is the whole.

> **Whole**
>
> Shape B:
> double hexagon

7. Cover Shape B with trapezoids.
   What fraction of the shape is covered by

   1 trapezoid? _____   2 trapezoids? _____   3 trapezoids? _____

8. Cover Shape B with rhombuses. What fraction of the shape is covered by

   1 rhombus? _____   3 rhombuses? _____   5 rhombuses? _____

9. Cover Shape B with triangles. What fraction of the shape is covered by

   1 triangle? _____   2 triangles? _____   3 triangles? _____

10. Cover Shape B with hexagons. What fraction of the shape is covered by

    1 hexagon? _____   2 hexagons? _____

11. Cover Shape B completely
    with 1 hexagon, 1 rhombus,
    1 triangle, and 1 trapezoid.
    Draw the result on the figure
    at the right. Label each part
    with a fraction.

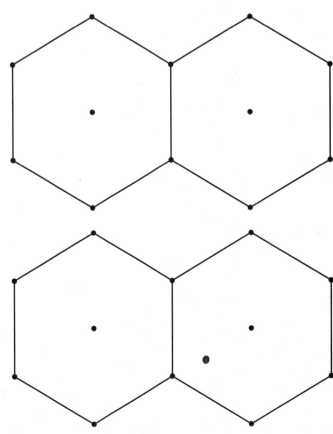

12. Cover Shape B completely
    with 1 trapezoid, 2 rhombuses,
    and 5 triangles. Draw the
    result on the figure at the right.
    Label each part with a fraction.

LESSON 7·4 Pattern-Block Fractions *continued*

Use *Math Masters,* page 212. For Problems 13–16, Shape C is the whole.

Whole

Shape C:
big hexagon

Try This

13. Cover Shape C with trapezoids.
What fraction of the shape is covered by

   1 trapezoid? _____   2 trapezoids? _____   6 trapezoids? _____

14. Cover Shape C with rhombuses. What fraction of the shape is covered by

   1 rhombus? _____   3 rhombuses? _____   6 rhombuses? _____

15. Cover Shape C with triangles. What fraction of the shape is covered by

   1 triangle? _____   3 triangles? _____   12 triangles? _____

16. Cover Shape C completely, using one or more trapezoids, rhombuses, triangles, and
hexagons. Draw the result on the big hexagon below. Label each part with a fraction.

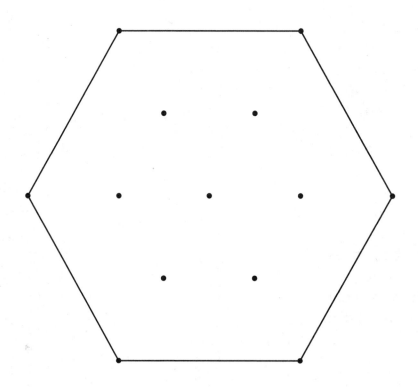

LESSON 7·4   **Math Boxes**

---

1. What fraction of the clock face is shaded?

_____

56

---

2. Draw angle *LMN* that measures 120°.

∠*LMN* is an _____
(acute or obtuse) angle.

93 142 143

---

3. a. In December, $\frac{3}{4}$ of a foot of snow fell on Wintersville. How many inches of snow fell?

_____ inches

b. Tina's daughter will be $\frac{5}{6}$ of a year old next week. How many months old will she be?

_____ months

59

---

4. Divide. Use a paper-and-pencil algorithm.

809 / 13 = _____

22 23 179

---

5. Each student eats an average of 17 servings of junk food per week. About how many servings of junk food would a class of 32 students eat in a week?

_____ servings

---

6. Find the area of the figure.

☐ = 1 square centimeter

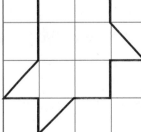

Area = _____ square cm

133

LESSON 7·5   **Fraction and Mixed-Number Sums & Differences**

1. Use pattern blocks to find fractions that add up to 1 whole. Draw lines to show the blocks you used. Write a number model to show that the sum of your fractions is 1.

Whole

hexagon

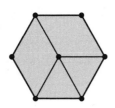      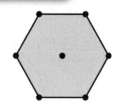

$\frac{1}{6} + \frac{1}{6} + \frac{1}{3} + \frac{1}{3} = 1$ _____   _____

      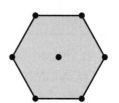

_____   _____   _____

2. Use pattern blocks to find fractions that add up to $\frac{2}{3}$. Draw lines to show the blocks you used. Write a number model to show that the sum of your fractions is $\frac{2}{3}$.

_____   _____   _____

Solve. You may use pattern blocks or any other method.

3. $\frac{3}{6} - \frac{2}{6} =$ _____

4. $\frac{2}{2} - \frac{1}{2} =$ _____

5. $\frac{2}{3} - \frac{1}{6} =$ _____

6. $\frac{5}{6} - \frac{1}{2} =$ _____

LESSON 7·5

Fraction and Mixed-Number Sums & Differences *cont.*

7. Use pattern blocks to find three different pairs of mixed numbers that add up to $2\frac{4}{6}$. Use your Geometry Template to illustrate the mixed numbers. Write a number model to show that the sum of each pair of mixed numbers is $2\frac{4}{6}$.

Whole
hexagon

a.

Number model: _____

b.

Number model: _____

c.

Number model: _____

Solve. You may use pattern blocks or any other method.

8. $2\frac{1}{2} - 1\frac{1}{2} =$ _____

9. $1\frac{2}{3} - 1\frac{1}{3} =$ _____

10. $1\frac{1}{6} - \frac{1}{3} =$ _____

11. $1\frac{1}{2} - \frac{5}{6} =$ _____

LESSON 7·5 **Solving Fraction Number Stories**

1. Rithik ate $\frac{3}{6}$ of a cheese pizza. He then ate $\frac{1}{6}$ of a veggie pizza.

   a. What fraction of a pizza did he eat in all? _____

      Number model: _____

   b. Did he eat more or less than a whole pizza? _____ How do you know?

      _____

2. Karina walked $\frac{1}{4}$ of a mile to school. After school, she walked $\frac{2}{4}$ of a mile to the store, and then $\frac{1}{4}$ of a mile back to her home.

   a. How far did she walk after school? _____

      Number model: _____

   b. How far did she walk in all? _____

      Number model: _____

3. Stephano is making pancakes and waffles for his guests.

   a. He needs $\frac{2}{3}$ cup of milk for the pancakes and $\frac{2}{3}$ cup of milk for the waffles. How much milk does he need in all? _____

      Number model: _____

   b. Stephano has $1\frac{2}{3}$ cups of milk. Will he have any left over? If so, how much milk will be left? _____

      Number model: _____

Try This

4. Kumba has one dollar. He spent $\frac{1}{2}$ of the dollar on a pencil and $\frac{2}{10}$ of the dollar on an eraser.

   a. What fraction of the dollar did he spend? _____

      Number model: _____

   b. What fraction of the dollar does he have left? _____

      Number model: _____

LESSON 7·5   **Math Boxes**

---

1. Circle $\frac{1}{5}$ of all the triangles. Mark Xs on $\frac{2}{3}$ of all the triangles.

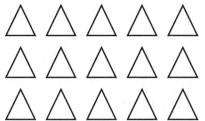

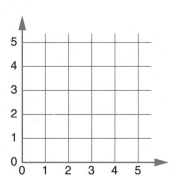

SRB
59

---

2. Insert parentheses to make these number sentences true.

a. $8.2 - 5.2 + 2.5 = 0.5$

b. $13.6 - 5 + 8 = 0.6$

c. $9.1 = 28.4 - 1.1 \div 3$

d. $9 * 2.5 + 3.5 = 54$

SRB
150

---

3. Plot and label each point on the coordinate grid.

A (5,0)

B (3,5)

C (1,4)

D (1,1)

E (2,4)

SRB
144

---

4. Draw and label a 45° angle.

This angle is an _____ (acute or obtuse) angle.

SRB
93 143

---

5. A bag contains

6 red blocks,
4 blue blocks,
7 green blocks, and
3 orange blocks.

You put your hand in the bag and, without looking, pull out a block. About what fraction of the time would you expect to get a blue block?

_____

SRB
45

---

6. If 1 centimeter on a map represents 10 kilometers, then

a. 6 cm represent _____ km.

b. 19.5 cm represent _____ km.

c. _____ cm represent 30 km.

d. _____ cm represent 55 km.

e. _____ cm represent 5 km.

SRB
145

LESSON 7·6   Math Boxes

1. Which fraction is another name for $\frac{6}{8}$?
   Fill in the circle next to the best answer.

   Ⓐ $\frac{1}{2}$

   Ⓑ $\frac{3}{4}$

   Ⓒ $\frac{4}{12}$

   Ⓓ $\frac{2}{4}$

   SRB 51

2. A bag contains

   2 blue blocks,
   3 red blocks,
   5 green blocks, and
   10 black blocks.

   You put your hand in the bag and, without looking, pull out a block. About what fraction of the time would you expect to get a black block?

   _____

   SRB 45

3. Use pattern blocks to help you solve these problems.

   a. $\frac{2}{6} + \frac{2}{6} =$ _____

   b. $\frac{1}{2} + \frac{1}{3} =$ _____

   c. $\frac{2}{3} - \frac{1}{3} =$ _____

   d. $\frac{2}{3} - \frac{1}{6} =$ _____

   SRB 55–57

4. ∠TAP is an _____ (acute or obtuse) angle.

   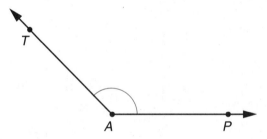

   The measure of ∠TAP is

   _____.

   SRB 93 142 143

5. Next month 486 students, teachers, and parents are going on a field trip to the zoo. Each bus holds 35 people. How many buses are needed for the trip?

   _____ buses

6. Tell if each of these is closest to 1 inch, 1 foot, or 1 yard.

   a. the length of your smile _____

   b. the length
   of your journal _____

   c. the distance from
   your waist to your feet _____

   d. the width of your wrist _____

   SRB 130

LESSON 7·6   **Adding and Subtracting Fractions and Mixed Numbers**

Use pattern blocks to help you solve the problems.

1. $\frac{1}{3} + \frac{1}{3} =$ _____

2. $\frac{4}{6} + \frac{2}{6} =$ _____

3. $\frac{2}{2} + \frac{1}{2} =$ _____

4. $\frac{1}{5} + \frac{3}{5} =$ _____

5. $1\frac{1}{4} + \frac{3}{4} =$ _____

6. $1\frac{2}{12} + 2\frac{3}{12} =$ _____

7. $1\frac{10}{10} + \frac{1}{10} =$ _____

8. $3\frac{4}{2} + 1\frac{1}{2} =$ _____

9. $\frac{5}{9} - \frac{2}{9} =$ _____

10. $\frac{2}{3} - \frac{1}{3} =$ _____

11. $\frac{6}{8} - \frac{2}{8} =$ _____

12. $\frac{3}{7} - \frac{2}{7} =$ _____

13. $1\frac{6}{9} - \frac{4}{9} =$ _____

14. $5\frac{2}{3} - 4\frac{1}{3} =$ _____

15. $10\frac{5}{6} - 6\frac{5}{6} =$ _____

16. $4\frac{4}{5} - 2\frac{5}{5} =$ _____

17. Ryan and Reggie baked an apple pie that was cut into 12 equal pieces. Ryan had $\frac{3}{12}$ of the pie, and Reggie ate $\frac{5}{12}$.

Who ate more? _____

What fraction of the pie did the boys eat all together? _____

18. Alice and Cherice run at the same park. On Saturday, Alice ran $\frac{3}{8}$ of a mile, and Cherice ran $\frac{7}{8}$ of a mile.

Who ran the shorter distance? _____

How far did Alice and Cherice run all together? _____ miles

LESSON 7·7   **Fraction and Mixed-Number Addition and Subtraction**

Add and subtract. Use pattern blocks to help you.

1. $\frac{1}{4} + \frac{2}{4} - \frac{1}{4} =$ _____

2. $\frac{3}{8} - \frac{2}{8} + \frac{4}{8} =$ _____

3. $\frac{4}{12} - \frac{2}{12} - \frac{1}{12} =$ _____

4. $7\frac{5}{6} + 1\frac{3}{6} - 1\frac{4}{6} =$ _____

5. $4\frac{3}{4} - \frac{2}{4} + 1\frac{1}{4} =$ _____

6. $\frac{4}{8} + \frac{2}{8} + 3\frac{2}{8} =$ _____

7. $\frac{2}{2} + \frac{2}{2} - \frac{1}{2} =$ _____

8. $\left(\frac{3}{8} - \frac{2}{8}\right) + \left(\frac{4}{8} - \frac{1}{8}\right) =$ _____

9. $\left(2\frac{4}{9} + \frac{2}{9}\right) - \left(\frac{1}{9} + 1\frac{3}{9}\right) =$ _____

10. $\left(10\frac{11}{12} + \frac{1}{12}\right) - \left(5\frac{3}{12} - \frac{1}{12}\right) =$ _____

11. $\left(\frac{6}{8} - \frac{1}{8}\right) - \left(\frac{4}{8} - \frac{2}{8}\right) =$ _____

12. $\left(\frac{3}{5} + \frac{1}{5}\right) + \left(\frac{2}{5} - \frac{1}{5}\right) =$ _____

13. Paulo, Regina, and Ted picked a bucket of apples on the field trip to the apple orchard. Paulo took $\frac{3}{16}$ of the apples, Regina took $\frac{6}{16}$ of the apples, and Ted took $\frac{1}{16}$ of the apples. They decided to give the rest of the apples to the teacher.

Who took the most apples? _____

What fraction of the apples did their teacher get? _____

How do you know?

_____

_____

_____

14. Julie was making a quilt. She had $\frac{5}{8}$ yard of fabric. She bought another $\frac{7}{8}$ yard of fabric. She gave $\frac{4}{8}$ yard of the fabric to her friend.

How many yards of fabric does she have left? _____ yard

LESSON 7·7   Many Names for Fractions

Color the squares and write the missing numerators.

Whole
square

1. Color $\frac{1}{2}$ of each large square.

   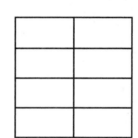   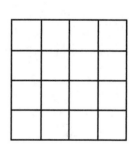

$\dfrac{1}{2}$ is colored. $\dfrac{}{4}$ is colored. $\dfrac{}{8}$ is colored.

2. Color $\frac{1}{4}$ of each large square.

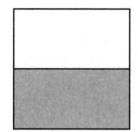   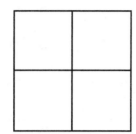   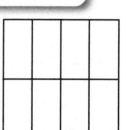

$\dfrac{}{4}$ is colored. $\dfrac{}{8}$ is colored. $\dfrac{}{16}$ is colored.

3. Color $\frac{3}{4}$ of each large square.

   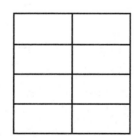   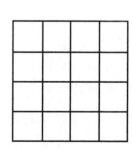

$\dfrac{}{4}$ is colored. $\dfrac{}{8}$ is colored. $\dfrac{}{16}$ is colored.

LESSON 7·7 **Math Boxes**

---

1. Circle $\frac{3}{8}$ of all the squares. Mark Xs on $\frac{1}{6}$ of all the squares.

☐ ☐ ☐ ☐ ☐ ☐
☐ ☐ ☐ ☐ ☐ ☐
☐ ☐ ☐ ☐ ☐ ☐
☐ ☐ ☐ ☐ ☐ ☐

SRB 59

2. Insert parentheses to make these number sentences true.

a. $2 * 3 + 10 = 26$

b. $12 = 6 * 6 - 4$

c. $24 - 5 * 2 = 38$

d. $12 + 24 = 3 * 6 + 6$

SRB 150

---

3. Plot and label each point on the coordinate grid.

A $(0,2)$

B $(4,0)$

C $(1,5)$

D $(5,5)$

E $(5,3)$

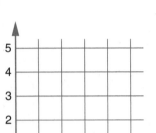

SRB 144

4. Draw and label a 125° angle.

This angle is an _____ (acute or obtuse) angle.

SRB 92 93 143

---

5. A bag contains

5 green blocks,
6 red blocks,
1 blue block, and
3 yellow blocks.

You put your hand in the bag and, without looking, pull out a block. About what fraction of the time would you expect to get a blue block?

_____

SRB 45

6. If 1 inch on a map represents 40 miles, then how many inches represent 10 miles? Fill in the circle next to the best answer.

Ⓐ 2 in.

Ⓑ $\frac{1}{4}$ in.

Ⓒ $\frac{1}{2}$ in.

Ⓓ 4 in.

SRB 145

Date _____ Time _____

Whole

large square

$\frac{1}{10}$, or 0.1

1.

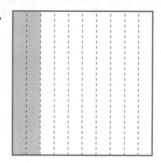

$\frac{2}{10}$ of the square is shaded.

How many tenths? _____

$\frac{2}{10} = 0.$_____

2.

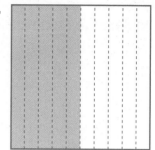

$\frac{1}{2}$ is shaded.

How many tenths? _____

$\frac{1}{2} = \dfrac{\boxed{}}{10} = 0.$_____

3.

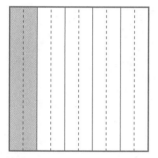

$\frac{1}{5}$ is shaded.

How many tenths? _____

$\frac{1}{5} = \dfrac{\boxed{}}{10} = 0.$_____

4.

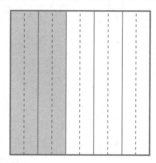

$\frac{2}{5}$ is shaded.

How many tenths? _____

$\frac{2}{5} = \dfrac{\boxed{}}{10} = 0.$_____

---

5. $\frac{3}{5} = \dfrac{\boxed{}}{10} = 0.$_____

6. $\frac{4}{5} = \dfrac{\boxed{}}{10} = 0.$_____

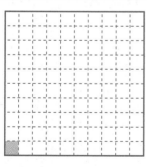

$\frac{1}{100}$, or 0.01

7.

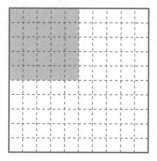

$\frac{1}{4}$ is shaded.

$\frac{1}{4} = \dfrac{\boxed{}}{100} = 0.$_____

8.

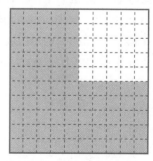

$\frac{3}{4}$ is shaded.

$\frac{3}{4} = \dfrac{\boxed{}}{100} = 0.$_____

LESSON 7·8   **Math Boxes**

---

1. Complete the name-collection box.

$$\frac{4}{5}$$

SRB 51

2. A bag contains

8 blue blocks,
2 red blocks,
1 green block, and
4 orange blocks.

You put your hand in the bag and, without looking, pull out a block. About what fraction of the time would you expect to get a red block?

_____

SRB 45

---

3. Use pattern blocks to help you solve these problems.

a. $\frac{1}{3} + \frac{1}{3} =$ _____

b. $\frac{2}{6} + \frac{2}{3} =$ _____

c. $\frac{5}{6} - \frac{1}{6} =$ _____

d. $\frac{4}{6} - \frac{1}{2} =$ _____

SRB 55–57

4. $\angle ART$ is an _____ (acute or obtuse) angle.

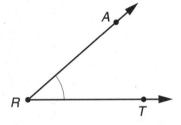

The measure of $\angle ART$ is

_____ °.

SRB 93 142 143

---

5. There are 252 pages in the book Ming is reading for his book report. He has two weeks to read the book. About how many pages should he read each day?

_____ pages

6. Tell if each of these is closest to 1 inch, 1 foot, or 1 yard.

a. the height of the door _____

b. the width of your journal _____

c. the length of your largest toe _____

d. the length of your shoe _____

SRB 130

LESSON 7·9 Comparing Fractions

Math Message: Eating Fractions

Quinn, Nancy, Diego, Paula, and Kiana were given 4 chocolate bars to share.
All 4 bars were the same size.

1. Quinn and Nancy shared a chocolate bar. Quinn ate $\frac{1}{4}$ of the bar, and Nancy ate $\frac{2}{4}$.

   Who ate more? _____

   How much of the bar was left? _____

2. Diego, Paula, and Kiana each ate part of the other chocolate bars. Diego ate $\frac{2}{3}$ of
   a bar, Paula ate $\frac{2}{5}$ of a bar, and Kiana ate $\frac{5}{6}$ of a bar.

   Who ate more, Diego or Paula? _____

   How do you know? _____

   _____

Comparing Fractions with $\frac{1}{2}$

Turn your Fraction Cards fraction-side up. Sort them into three piles:

   ◆ fractions less than $\frac{1}{2}$

   ◆ fractions equal to $\frac{1}{2}$

   ◆ fractions greater than $\frac{1}{2}$

You can turn the cards over to check your work. When you are finished,
write the fractions in each pile in the correct box below.

Less than $\frac{1}{2}$	Equal to $\frac{1}{2}$	Greater than $\frac{1}{2}$

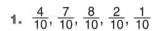 **LESSON 7·9** **Ordering Fractions**

Write the fractions in order from smallest to largest.

1. $\frac{4}{10}$, $\frac{7}{10}$, $\frac{8}{10}$, $\frac{2}{10}$, $\frac{1}{10}$

_____ _____ _____ _____ _____
smallest   **largest**

2. $\frac{1}{4}$, $\frac{1}{2}$, $\frac{1}{9}$, $\frac{1}{5}$, $\frac{1}{100}$

_____ _____ _____ _____ _____
smallest   **largest**

3. $\frac{2}{4}$, $\frac{2}{2}$, $\frac{2}{9}$, $\frac{2}{5}$, $\frac{2}{100}$

_____ _____ _____ _____ _____
smallest   **largest**

4. $\frac{4}{25}$, $\frac{1}{25}$, $\frac{7}{8}$, $\frac{6}{12}$, $\frac{7}{15}$

_____ _____ _____ _____ _____
smallest   **largest**

5. Choose 5 fractions or mixed numbers. Write them in order from smallest to largest.

_____ _____ _____ _____ _____
smallest   **largest**

6. Which fraction is larger: $\frac{2}{5}$ or $\frac{2}{7}$? _____ Explain how you know.

_____

_____

_____

LESSON 7·9   **Math Boxes**

---

1. Sari spends $\frac{1}{3}$ of the day at school. Lunch, recess, music, gym, and art make up $\frac{1}{4}$ of her total time at school. How many hours are spent at these activities?

   _____ hours

   Show how you solved this problem.

   _____

   _____

   SRB
   59

2. Multiply. Use a paper-and-pencil algorithm.

   _____ = 92 * 56

   SRB
   18 19

---

3. Adena drew a line segment $\frac{3}{4}$ inch long. Then she erased $\frac{1}{2}$ inch. How long is the line segment now? Fill in the circle next to the best answer.

   Ⓐ $\frac{4}{6}$ in.

   Ⓑ $\frac{2}{2}$ in.

   Ⓒ $\frac{1}{4}$ in.

   Ⓓ $1\frac{1}{4}$ in.

   SRB
   55–57

4. Write an equivalent fraction, decimal, or whole number.

Decimal	Fraction
a. 0.40	_____
b. _____	$\frac{3}{10}$
c. _____	$\frac{100}{100}$
d. 0.6	_____

   SRB
   61 62

---

5. Complete the table and write the rule.

   Rule: _____

in	out
6.19	11.92
12.03	
	8.99
	5.74
4.41	10.14

   SRB
   162–166

6. Complete.

   a. 17 in. = _____ ft _____ in.

   b. 43 in. = _____ ft _____ in.

   c. 6 ft = _____ yd

   d. 11 ft = _____ yd _____ ft

   e. 73 yd = _____ ft

   SRB
   129

LESSON 7·10   # What Is the ONE?

Math Message

1. If the triangle below is $\frac{1}{3}$, then what is the whole—the ONE? Draw it on the grid.

2. If $\frac{1}{4}$ of Mrs. Chin's class is 8 students, then
   how many students does she have altogether? _____ students

Use your Geometry Template to draw the answers for Problems 3–6.

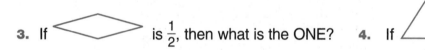

3. If ⬦ is $\frac{1}{2}$, then what is the ONE?   4. If ▱ is $\frac{1}{4}$, then what is the ONE?

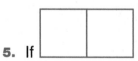

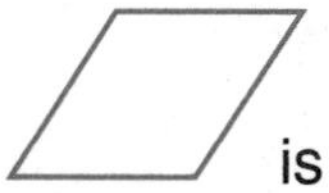

5. If � rectangle is $\frac{2}{3}$, then what is the ONE?   6. If ⬡ is $\frac{2}{5}$, then what is the ONE?

**LESSON
7·10**   ## What is the ONE? *continued*

Solve. If you wish, draw pictures at the bottom of the page to help you
solve the problems.

7. If ◯◯◯◯◯ is $\frac{1}{3}$, then what is the ONE?   _____ counters

8. If ◯◯◯◯ is $\frac{1}{4}$, then what is the ONE?   _____ counters

9. If 10 counters are $\frac{2}{5}$, then what is the ONE?   _____ counters

10. If 12 counters are $\frac{3}{4}$, then what is the ONE?   _____ counters

11. If $\frac{1}{5}$ of the cookies that Mrs. Jackson baked is 12,
    then how many cookies did she bake in all?   _____ cookies

12. In Mr. Mendez's class, $\frac{3}{4}$ of the students take music
    lessons. That is, 15 students take music lessons.
    How many students are in Mr. Mendez's class?   _____ students

13. Explain how you solved Problem 12.

    _____

    _____

    _____

LESSON 7·10   Insect Data

Veronica collected 15 insects for a science project. She measured the length of each insect to the nearest $\frac{1}{8}$ inch. Her measurements are shown in the table below.

Insect	Length (to the nearest $\frac{1}{8}$ inch)	Insect	Length (to the nearest $\frac{1}{8}$ inch)
Darner dragonfly	$1\frac{1}{2}$	Red legged grasshopper	$1\frac{1}{8}$
Boreal firefly	$\frac{3}{8}$	American cockroach	$1\frac{1}{2}$
Yellow bumblebee	$\frac{3}{4}$	June beetle	$\frac{5}{8}$
Damselfly	$1\frac{1}{4}$	Paper wasp	$\frac{7}{8}$
Ground beetle	$\frac{7}{8}$	Field cricket	$\frac{7}{8}$
Green lacewing	1	Indian meal moth	$\frac{3}{8}$
Lady bug	$\frac{5}{8}$	Katydid	$1\frac{3}{4}$
Carolina mantid	$1\frac{3}{4}$		

Plot the insect lengths on the line plot below. Then use the completed plot to answer the questions on the next page.

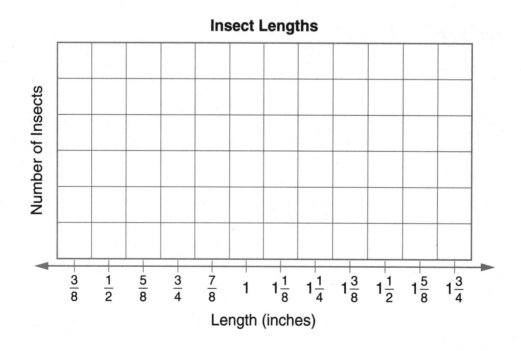

Insect Lengths

LESSON 7·10 Insect Data *continued*

Use the line plot on journal page 209A to answer the questions. Write a number model to summarize each problem.

1. a. What is the maximum insect length? _____ in. The minimum? _____ in.

   b. What is the range of the data set? _____ in.   Number model: _____

2. a. What is the median of the data set? _____ in.

   b. How much longer is the median length than the minimum length? _____ in.

   Number model: _____

3. a. What is the mode of the data set? _____ in.

   b. How much longer is the maximum length than the mode length? _____ in.

   Number model: _____

4. Two insects have the maximum length. What is the difference in length between these insects and the next-longest insects? _____ in.

   Number model: _____

5. There are three insects in Veronica's collection that are from $\frac{1}{2}$ inch to $\frac{3}{4}$ inch long. If these three insects were placed end to end, how long would the line of insects be?

   _____ in.   Number model: _____

6. How long would the line of insects be if all the insects less than $\frac{1}{2}$ inch long were placed end to end? _____ in.

   Number model: _____

7. Make up and solve your own problem about the insect data.

   _____

   _____

   Number model: _____

Math Boxes

---

1. Name the shaded area as a fraction and a decimal.

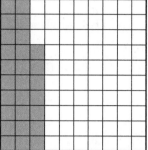

a. fraction:

_____

b. decimal:

_____

27 61

---

2. Which number sentence is true? Fill in the circle next to the best answer.

Ⓐ $\frac{5}{6} < \frac{1}{6}$

Ⓑ $\frac{4}{10} > \frac{4}{5}$

Ⓒ $\frac{1}{7} > \frac{1}{100}$

Ⓓ $\frac{2}{12} = \frac{3}{6}$

53 54

---

3. Write 6 fractions equivalent to $\frac{14}{16}$.

_____ _____

_____ _____

_____ _____

49–51

---

4. Divide. Use a paper-and-pencil algorithm.

$\frac{723}{14} =$ _____

22 23
179

---

5. Multiply. Use a paper-and-pencil algorithm.

_____ = 68 * 124

18 19

---

6. Compare.

a. 1 day is _____ times as long as 2 hours.

b. 6 years is _____ times as long as 4 months.

c. 3 gallons is _____ times as much as 8 cups.

d. 8 cm is _____ times as long as 2 mm.

e. 1 meter is _____ times as long as 2 cm.

315

LESSON 7·11   **Making Spinners**

1. Make a spinner. Color the circle in 6 different colors. Design the spinner so that the paper clip has the **same chance** of landing on each of the colors.

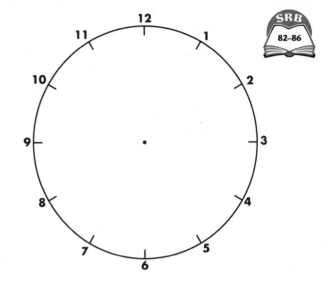

2. Make another spinner. Color the circle red, blue, and green so that the paper clip has

   ◆ a $\frac{1}{6}$ chance of landing on red

   and

   ◆ a $\frac{1}{3}$ chance of landing on blue.

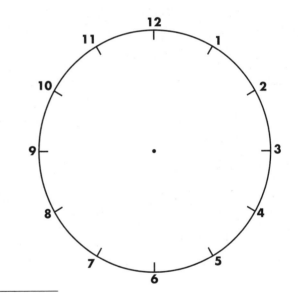

   a. What fraction of the circle did you color

      red? _____   blue? _____   green? _____

   b. Suppose you plan to spin the paper clip 24 times. About how many times would you expect it to land on

      red? _____   blue? _____   green? _____

   c. Suppose you plan to spin the paper clip 90 times. About how many times would you expect it to land on

      red? _____   blue? _____   green? _____

LESSON 7·11 **Math Boxes**

---

1. According to a survey of 800 students at Martin Elementary, about $\frac{3}{4}$ of them chose pizza as their favorite food. Of those who chose pizza, $\frac{1}{2}$ liked pepperoni topping the best. How many students liked pepperoni topping the best?

_____ students

SRB 59

---

2. Multiply. Use a paper-and-pencil algorithm.

71 * 38 = _____

SRB 18 19

---

3. a. Hannah drew a line segment $1\frac{5}{8}$ inches long. Then she erased $\frac{1}{2}$ inch. How long is the line segment now?

_____ inches

b. Joshua drew a line segment $\frac{7}{8}$ inch long. Then he added another $\frac{3}{4}$ inch. How long is the line segment now?

_____ inches

SRB 55–57

---

4. Write an equivalent fraction, decimal, or whole number.

	Decimal	Fraction
a.	0.70	_____
b.	_____	$\frac{25}{100}$
c.	_____	$\frac{9}{9}$
d.	0.2	_____

SRB 61 62

---

5. Complete the table and write the rule.

Rule: _____

in	out
100.54	97.58
	52.95
72.03	
	67.44
59.21	56.25

SRB 162–166

---

6. Complete.

a. 5 ft = _____ yd _____ ft

b. 40 in. = _____ ft _____ in.

c. 80 in. = _____ yd _____ in.

d. 108 in. = _____ ft

e. $\frac{1}{3}$ yd = _____ in.

SRB 129

LESSON 7·12 **Expected Spinner Results**

1. If this spinner is spun 24 times, how many times do you expect it to land on each color?

   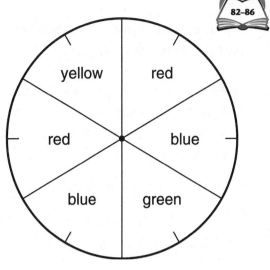

   a. Fill in the table.

Color	Expected Number in 24 Spins
red	
blue	
yellow	
green	
Total	**24**

   b. Explain how you determined the expected number of times the spinner would land on each color.

   _____

   _____

   _____

   _____

   Try This

2. If a six-sided die is rolled 12 times, how many times would you expect to roll

   a. an odd number? _____

   b. a number less than 4? _____

   c. a 6? _____

   d. a square number? _____

   e. a triangular number? _____

   f. a prime number? _____

LESSON 7·12 **A Cube-Drop Experiment**

Getting Ready

1. Follow the directions for coloring the grid on *Math Masters,* page 238. You may color the squares in any way. The colors can even form a pattern or a picture.

2. For this experiment, you are going to place your grid on the floor and hold a centimeter cube about 2 feet above the grid. Without aiming, you will let it drop onto the grid. You will then record the color of the square on which the cube finally lands.

   ◆ If the cube does not land on the grid, the drop does not count.

   ◆ If the cube lands on more than one color, record the color that is covered by most of the cube. If you cannot tell, the toss does not count.

Making a Prediction

3. On which color is the cube *most likely* to land? _____

4. On which color is it *least likely* to land? _____

5. Suppose you were to drop the cube 100 times. How many times would you expect it to land on each color? Record your predictions below.

Predicted Results of 100 Cube Drops

Color	Number of Squares	Predicted Results Fraction	Percent
yellow	1	$\frac{1}{100}$	_____%
red	4		_____%
green	10		_____%
blue	35		_____%
white	50		_____%
Total	**100**	**1**	**100%**

LESSON 7·12

A Cube-Drop Experiment *continued*

Doing the Experiment

You and your partner will each drop a centimeter cube onto your own colored grid.

6. One partner drops the cube. The other partner records the color in the grid below by writing a letter in one of the squares. Drop the cube a total of 50 times.

Write
y for yellow,
r for red,
g for green,
b for blue, and
w for white.

7. Then trade roles. Do another 50 drops, and record the results in the other partner's journal.

8. Count the number for each color.

   Write it in the "Number of Drops" column.

   Check that the total is 50.

My Results for 50 Cube Drops		
Color	**Number of Drops**	**Percent**
yellow		
red		
green		
blue		
white		
Total	**50**	**100%**

9. When you have finished, fill in the percent column in the table.

   Example: If your cube landed on blue 15 times out of 50 drops, this is the same as 30 times out of 100 drops, or 30% of the time.

LESSON 7·12 Place Value in Whole Numbers

SRB
4

1. Write these numbers in order from least to greatest.

   964   9,460   96,400   400,960   94,600

   _____

   _____

   _____

   _____

   _____

2. A number has

   5 in the hundreds place,
   7 in the ten-thousands place,
   0 in the ones place,
   9 in the thousands place, and
   8 in the tens place.

   Write the number.

   _____ _____ , _____ _____ _____

3. Write the greatest number you can make with the following digits:

   3   5   0   7   9   2

   _____

4. What is the value of the digit 8 in the numerals below?

   a. **8**07,941 _____

   b. 5**8**3 _____

   c. **8**,714 _____

   d. **8**6,490 _____

5. Write each number using digits.

   a. four hundred eighty-seven thousand, sixty-three

   _____

   b. fifteen thousand, two hundred ninety-seven

   _____

6. I am a 5-digit number.

   ◆ The digit in the thousands place is the result of dividing 64 by 8.

   ◆ The digit in the ones place is the result of dividing 63 by 9.

   ◆ The digit in the ten-thousands place is the result of dividing 54 by 6.

   ◆ The digit in the tens place is the result of dividing 40 by 5.

   ◆ The digit in the hundreds place is the result of dividing 33 by 11.

   What number am I?

   _____ _____ , _____ _____ _____

LESSON 7·12   Math Boxes

---

1. Name the shaded area as a fraction and a decimal.

   a. fraction:

      _____

   b. decimal:

      _____

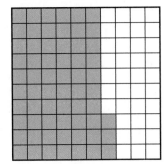

SRB 27 61

2. Write <, >, or = to make each number sentence true.

   a. $\dfrac{3}{8}$ _____ $\dfrac{7}{8}$

   b. $\dfrac{5}{12}$ _____ $\dfrac{5}{6}$

   c. $\dfrac{1}{4}$ _____ $\dfrac{1}{15}$

   d. $\dfrac{500}{1000}$ _____ $\dfrac{8}{16}$

   e. $\dfrac{6}{7}$ _____ $\dfrac{19}{20}$

SRB 53 54

---

3. Write 6 fractions equivalent to $\dfrac{1}{6}$.

   _____   _____

   _____   _____

   _____   _____

SRB 49–51

4. Divide. Use a paper-and-pencil algorithm.

   $\dfrac{769}{15}$ = _____

SRB 22 23 179

---

5. Multiply. Use a paper-and-pencil algorithm.

   _____ = 46 * 206

SRB 18 19

6. Compare.

   a. 1 day is _____ times as long as 6 hours.

   b. 6 years is _____ times as long as 2 months.

   c. 3 gallons is _____ times as much as 4 cups.

   d. 8 cm is _____ times as long as 5 mm.

   e. 1 meter is _____ times as long as 10 cm.

SRB 315

LESSON 7·12a   Multiples of Unit Fractions

For Problems 1–3, fill in the blanks to complete an equation describing the number line.

1.
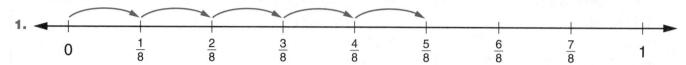

Equation: 5 * _____ = _____

2.
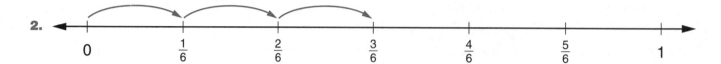

Equation: _____ * $\frac{1}{6}$ = _____

3.
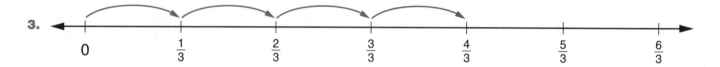

Equation: _____ * _____ = _____

For Problems 4–6, use the number line to help you multiply the fraction by the whole number.

4.
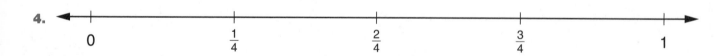

Equation: 2 * $\frac{1}{4}$ = _____

5.

Equation: 6 * $\frac{1}{10}$ = _____

6.
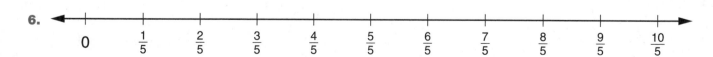

Equation: 7 * $\frac{1}{5}$ = _____

Example 1: Equation: $6 * \frac{1}{5} = \frac{6}{5}$

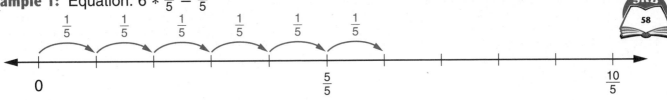

Example 2: Equation: $3 * \frac{2}{5} = \frac{6}{5}$

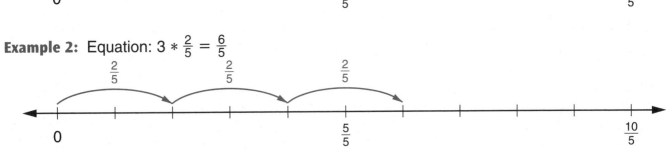

Write an equation to describe each number line.

1. a.

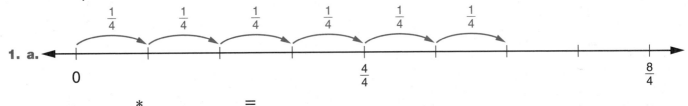

_____ * _____ = _____

b.

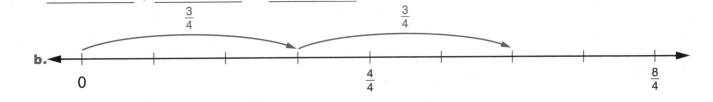

_____ * _____ = _____

2. a.

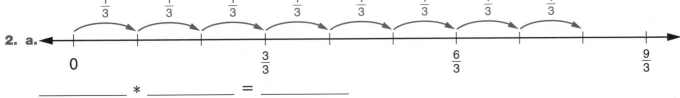

_____ * _____ = _____

b.

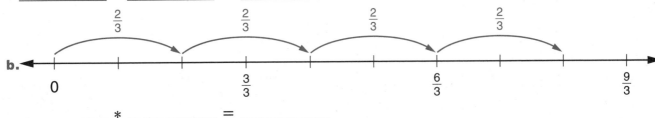

_____ * _____ = _____

3. Study the pairs of number lines above. Use the patterns you see to describe a way to multiply a fraction by a whole number.

_____

_____

LESSON
7·12a   **Multiplying Fractions by Whole Numbers**
SRB
58

Use number lines to help you solve the problems.

1. $5 * \frac{1}{6} =$ _____

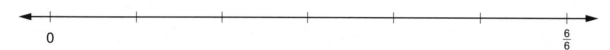

0 $\frac{6}{6}$

2. $6 * \frac{1}{3} =$ _____

0                        $\frac{3}{3}$ $\frac{6}{3}$ $\frac{9}{3}$

3. _____ $= 3 * \frac{1}{8}$

4. $2 * \frac{4}{3} =$ _____

0                        $\frac{3}{3}$ $\frac{6}{3}$ $\frac{9}{3}$

5. _____ $= 4 * \frac{3}{8}$

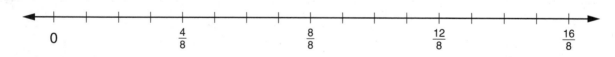

0                    $\frac{4}{8}$ $\frac{8}{8}$ $\frac{12}{8}$ $\frac{16}{8}$

6. _____ $= 3 * \frac{2}{10}$

LESSON 7·12a   Solving Number Stories

Suma and her sister Puja are making 12 blueberry-wheat muffins for breakfast. The recipe lists the following ingredients:

1 cup flour	1 egg
$\frac{1}{2}$ cup whole-wheat flour	$\frac{1}{2}$ cup skim milk
2 teaspoons baking powder	$\frac{2}{3}$ cup honey
$\frac{3}{4}$ cup blueberries	$\frac{1}{4}$ cup cooking oil
$\frac{1}{4}$ teaspoon salt	$\frac{3}{8}$ teaspoon cinnamon

Use the list of recipe ingredients to help you solve the number stories below. For each problem, write an equation to show what you did.

1. The sisters decided to double the recipe.

   a. How many cups of whole-wheat flour do they need now?

   _____ cup(s)   Equation: _____

   b. How many cups of blueberries do they need now?

   _____ cup(s)   Equation: _____

   c. How many cups of honey do they need now?

   _____ cup(s)   Equation: _____

2. Suma and Puja decide to make 48 muffins instead of 12.

   a. How many teaspoons of salt do they need now?

   _____ teaspoon(s)   Equation: _____

   b. How many teaspoons of cinnamon do they need now?

   _____ teaspoon(s)   Equation: _____

   c. How many cups of skim milk do they need now?

   _____ cup(s)   Equation: _____

LESSON 7·12a

Solving Number Stories *continued*

The Hillside Elementary School walking club meets every Monday after school. The table below shows how far some students walked at their last meeting.

Student	Katie	Mahpara	Nikhil	Cole	Maria	Jack
Miles	$\frac{1}{3}$	$\frac{9}{10}$	$\frac{5}{4}$	$\frac{5}{2}$	$\frac{4}{3}$	$\frac{5}{6}$

Use the information in the table to solve the number stories.

3. a. If Katie walks the same distance at every meeting, how far will she walk after 2 meetings? _____ miles

   b. After 7 meetings? _____ miles

   c. After 7 meetings, Katie will have walked between _____. Circle the best answer.

         1 and 2 miles   2 and 3 miles   3 and 4 miles

4. a. If Jack walks the same distance at every meeting, how far will he walk after 3 meetings? _____ miles

   b. After 3 meetings, Jack will have walked between _____. Circle the best answer.

         1 and 2 miles   2 and 3 miles   3 and 4 miles

5. a. If Mahpara walks the same distance at every meeting, how far will she walk after 4 meetings? _____ miles

   b. After 4 meetings, Mahpara will have walked between _____. Circle the best answer.

         1 and 2 miles   2 and 3 miles   3 and 4 miles

Try This

6. If Cole walks the same distance at every meeting and wants to walk a total of $\frac{15}{2}$ miles, how many meetings will he need to attend? _____ meetings

7. Make up your own multiplication number story about Nikhil or Maria.

   _____

   _____

217E

LESSON 7·12a **Math Boxes**

---

1. Karen used 60 square feet of her back yard for a garden. Vegetables fill $\frac{3}{5}$ of her garden space. Tomato plants fill $\frac{1}{6}$ of the space taken up by vegetables. How many square feet are used for tomatoes?

_____ square feet

SRB
59

---

2. Multiply. Use a paper-and-pencil algorithm.

_____ = 87 * 43

SRB
18 19

---

3. a. Lukasz drew a line segment that was $2\frac{2}{8}$ inches long. Then he extended it another $2\frac{3}{8}$ inches. How long is the line segment now?

_____ inches

b. Sybil drew a line segment $3\frac{1}{8}$ inches long. Then she extended it another $2\frac{3}{4}$ inches. How long is the line segment now?

_____ inches

SRB
55–57

---

4. Write an equivalent fraction, decimal, or whole number.

Decimal	Fraction
a. 0.60	_____
b. _____	$\frac{65}{100}$
c. _____	$\frac{50}{50}$
d. 0.9	_____

SRB
61 62

---

5. Complete the table and write the rule.

Rule: _____

in	out
104.16	100.67
	83.86
45.72	
55.41	
77.69	74.20

SRB
162–166

---

6. Complete.

a. 42 in. = _____ ft _____ in.

b. 16 ft = _____ in.

c. 67 in. = _____ ft _____ in.

d. 22 ft = _____ yd _____ ft

e. $1\frac{1}{2}$ yd = _____ ft _____ in.

SRB
129

LESSON 7·13   **Math Boxes**

---

1. Measure the length and width of your desk to the nearest half-inch. Find its perimeter.

 a. Length = _____ inches

 b. Width = _____ inches

 c. Perimeter = _____ inches

SRB 131

---

2. Find the area of the figure.

☐ = 1 square centimeter

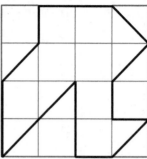

Area = _____ square cm

SRB 133

---

3. If 1 centimeter on a map represents 20 kilometers, then

 a. 8 cm represent _____ km.

 b. 3.5 cm represent _____ km.

 c. _____ cm represent 30 km.

 d. _____ cm represent 50 km.

 e. _____ cm represents 10 km.

SRB 145

---

4. Tell if each of these is closest to 1 inch, 1 foot, or 1 yard.

 a. the width of a door _____

 b. the width of your ankle _____

 c. the length of your little finger _____

 d. the length of your forearm _____

SRB 130

---

5. Complete.

 a. 26 in. = _____ ft _____ in.

 b. 57 in. = _____ ft _____ in.

 c. 9 ft = _____ yd

 d. 16 ft = _____ yd _____ ft

 e. 8 yd = _____ ft

SRB 129

---

6. Compare.

 a. 1 day is _____ times as long as 12 hours.

 b. 3 years is _____ times as long as 6 months.

 c. 12 cm is _____ times as long as 2 mm.

 d. 1 m is _____ times as long as 20 cm.

 e. 3 gallons is _____ times as much as 2 cups.

SRB 315

LESSON 8·1   Kitchen Layouts and Kitchen Efficiency

Here are four common ways to arrange the appliances in a kitchen:

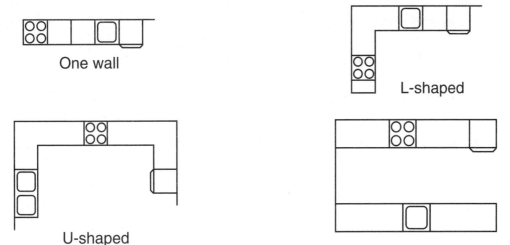

One wall

L-shaped

U-shaped

Pullman or galley

Pullman kitchens are usually found on passenger trains. **Galleys** are the kitchens on boats and airplanes. The kitchen areas on trains, boats, and airplanes are small. The cooking area is usually lined up against a single wall (a one-wall kitchen) or against two walls with a corridor between them (a Pullman or galley kitchen).

◆ What kind of kitchen layout do you have in your home? Circle one.

    One wall   L-shaped   U-shaped   Pullman or galley

Kitchen efficiency experts are people who study the ways we use our kitchens. They carry out **time-and-motion** studies to find how long it takes to do some kitchen tasks and how much a person has to move about in order to do them. They want to find the best ways to arrange the stove, the sink, and the refrigerator. In an efficient kitchen, a person should have to do very little walking to move from one appliance to another. However, the appliances should not be too close to each other, because the person would feel cramped.

A bird's-eye sketch is often drawn to see how well the appliances in a kitchen are arranged. The stove, the sink, and the refrigerator are connected with line segments as shown below. These segments form a triangle called a **work triangle.** The work triangle shows the distance between pairs of appliances.

Work Triangle

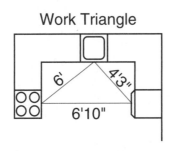

6'   4'3"

6'10"

LESSON 8·1 **Layout of My Kitchen**

1. Copy the distances between your appliances from *Math Masters,* page 235.

   Between stove and refrigerator: About _____ feet _____ inches

   Between refrigerator and sink: About _____ feet _____ inches

   Between sink and stove: About _____ feet _____ inches

2. Cut out the sketch of your kitchen from *Math Masters,* page 236 and tape it in the space below.

Date _____ Time _____

LESSON 8·1 How Efficient Is My Kitchen?

Answer the questions below to see how well the appliances in your kitchen are arranged.

1. With a straightedge, draw a triangle connecting the appliances in your sketch on page 220. Write the distances between the appliances on the sides of your triangle. This triangle is called a **work triangle.**

2. Find the **perimeter** of your work triangle. Show your work.

   _____ feet _____ inches

   _____ feet _____ inches

   + _____ feet _____ inches

   _____

   The perimeter is about _____ feet _____ inches.

   That's close to _____ feet.

3. Kitchen efficiency experts recommend the following distances between appliances:

   Between stove and refrigerator: 4 feet to 9 feet

   Between refrigerator and sink: 4 feet to 7 feet

   Between sink and stove: 4 feet to 6 feet

   Does your kitchen meet these recommendations? _____

4. How many students reported their work triangle perimeters? _____ students

   The minimum perimeter is about _____ feet.

   The maximum perimeter is about _____ feet.

   The mode of the perimeters is about _____ feet.

   The median perimeter is about _____ feet.

LESSON 8·1   Work Triangles

1. a. Below, draw a work triangle that meets all of the following conditions:

   ◆ The perimeter is 21 feet.

   ◆ The length of each side is a whole number of feet.

   ◆ The length of each side is in the recommended range:

        Between stove and refrigerator: 4 feet to 9 feet

        Between refrigerator and sink: 4 feet to 7 feet

        Between sink and stove: 4 feet to 6 feet

    b. Write the distances on the sides of your triangle.

    c. Label each vertex (corner) of the triangle as *stove, sink,* or *refrigerator.*

2. Below, draw a different work triangle that meets the same conditions listed in Problem 1.

LESSON 8·1 | **Math Boxes**

---

1. Some fourth graders were asked how many minutes they spend studying at home per week. Here are the responses from ten students:

130, 45, 240, 35, 160, 185, 120, 20, 55, 160

a. What is the mode? _____ minutes

b. What is the median? _____ minutes

---

2. Insert >, <, or = to make each number sentence true.

a. $\frac{11}{12}$ _____ $\frac{19}{20}$

b. $\frac{1}{4}$ _____ $\frac{1}{9}$

c. $\frac{4}{9}$ _____ $\frac{12}{27}$

d. $\frac{10}{12}$ _____ $\frac{30}{36}$

e. $\frac{7}{2}$ _____ $\frac{21}{6}$

---

3. a. Use your Geometry Template to draw an equilateral triangle.

b. Measure one of the angles with your protractor. Record the measure.

_____ °

---

4. If you spin the spinner below 100 times, how many times would you expect it to land

on red? _____ times

on black? _____ times

on white? _____ times

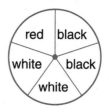

---

5. Solve the open sentence $\frac{1}{4} + y = \frac{3}{8}$. Circle the best answer.

A $y = \frac{2}{4}$

B $y = \frac{4}{12}$

C $y = \frac{1}{8}$

D $y = \frac{1}{4}$

---

6. A store is giving a 50% discount on all merchandise. Find the discounted prices.

Regular price	Discounted price
$26.00	_____
$0.48	_____
$140.60	_____
$65.24	_____

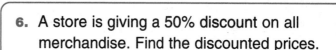

LESSON 8·2

A Floor Plan of My Classroom

When architects design a room or house, they usually make two drawings. The first drawing is called a **rough floor plan.** It is not carefully drawn. But the rough floor plan includes all of the information that is needed to make an accurate drawing. The second drawing is called a **scale drawing.** It is drawn on a grid and is very accurate.

Rough floor plan for a bedroom

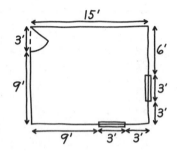

Scale drawing for a bedroom
(1 grid length represents 1 foot.)

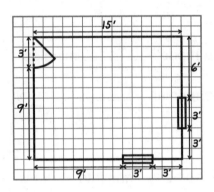

1. What information do you need to draw a rough floor plan?

_____

_____

_____

Architects use these symbols to show windows and doors:

window   door opening to left   door opening to right

2. Make a rough floor plan of your classroom in the space below.

LESSON 8·2 **A Floor Plan of My Classroom** *continued*

3. Make a scale drawing of your classroom. Scale: _____ inch represents _____ foot.

   Each side of a small square in the grid below is $\frac{1}{4}$ inch long.

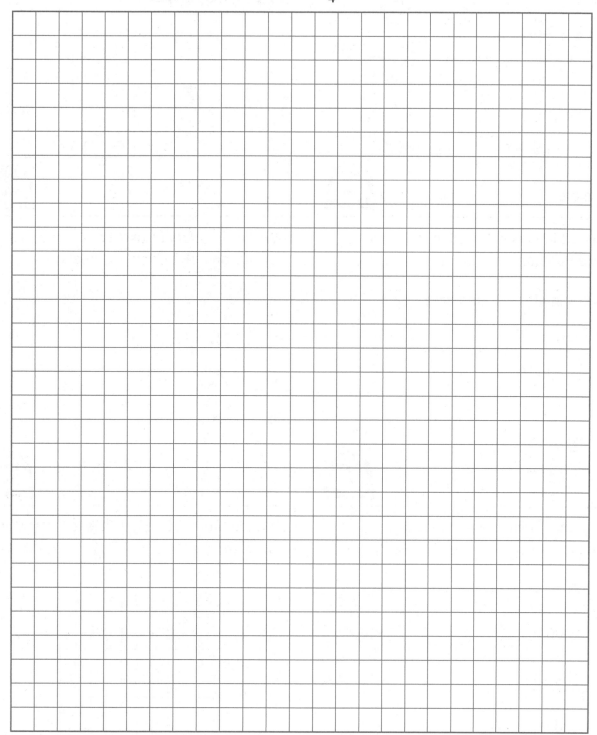

For use in Lesson 8-3: The area of my classroom is about _____ square feet.

LESSON 8·2   **Math Boxes**

---

1. Measure the sides of the figure to the nearest centimeter. Then find its perimeter.

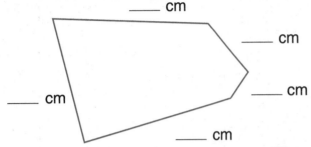

_____ cm

_____ cm

_____ cm

_____ cm

_____ cm

Perimeter = _____ cm

SRB 131

---

2. If you tossed a coin onto the grid below, about what fraction of the time would you expect it to land on R?

_____

R	O	P	E
O	P	E	R
P	E	R	O
E	R	O	P

SRB 45 84

---

3. Write an equivalent fraction, decimal, or whole number.

	Decimal	Fraction
a.	0.8	_____
b.	_____	$\frac{65}{100}$
c.	_____	$\frac{15}{15}$
d.	0.90	_____

SRB 61

---

4. Which number is closest to the product of 510 and 18? Circle the best answer.

  A   100

  B   1,000

  C   10,000

  D   100,000

SRB 181

---

5. Write each number in exponential notation.

  a.   100 = _____

  b.   10,000 = _____

  c.   1,000,000 = _____

  d.   1,000 = _____

SRB 5

---

6. Shade more than $\frac{2}{100}$ but less than $\frac{1}{10}$ of the grid.

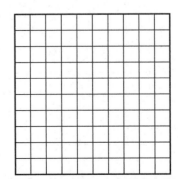

SRB 27

Date _____ Time _____

Write an equation to describe each number line.

1.

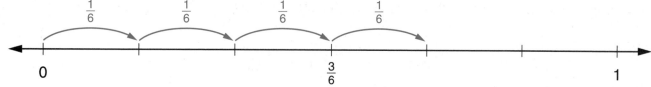

_____ * _____ = _____

2.

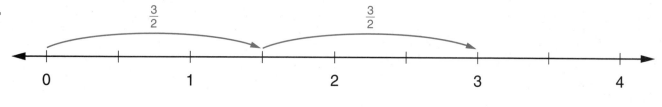

_____ * _____ = _____

Use number lines to help you solve the problems.

3. _____ = $2 * \frac{3}{5}$

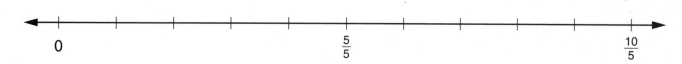

4. $3 * \frac{2}{7} =$ _____

Solve. You may use a visual fraction model such as a number line or any other method.

5. $7 * \frac{1}{12} =$ _____   **6.** _____ $= 5 * \frac{3}{10}$

7. _____ $= 3 * \frac{4}{7}$ **8.** $2 * \frac{5}{9} =$ _____

LESSON 8·2 **Solving Number Stories**

SRB
58

The students in the teen living class at Eagle Ridge Middle School are sewing baggy shorts for a fundraiser. They plan to sell each pair for $7.50. Use the information in the table to solve the number stories.

Size	Waist (in.)	Fabric (yd)
S	24–30	$\frac{7}{8}$
M	31–33	1
L	34–36	1
XL	37–40	$\frac{9}{8}$
XXL	41–45	$\frac{5}{4}$

1. **a.** How much fabric will Kent need if he wants to sew 4 pairs of S shorts?

   _____ yards   Equation: _____

   b. Kent needs between _____ yards of fabric. Circle the best answer.

   1 and 2   2 and 3   3 and 4

2. **a.** Monique wants to sew 3 pairs of XXL shorts. How much fabric will she need?

   _____ yards   Equation: _____

   b. Monique needs between _____ yards of fabric. Circle the best answer.

   1 and 2   2 and 3   3 and 4

3. **a.** Omar wants to sew 2 pairs of shorts that will fit a person with a 38-inch waist. How much fabric will he need?

   _____ yards   Equation: _____

   b. Omar needs between _____ yards of fabric. Circle the best answer.

   1 and 2   2 and 3   3 and 4

4. If Olivia has $\frac{21}{8}$ yards of fabric, how many pairs of S shorts will she be able to sew?

   _____ pairs

5. Ryan sewed $30.00 worth of XXL shorts. How many yards of fabric did he use?

   _____ yards

LESSON 8·3 **Areas of Polygons**

Find the area of each polygon.

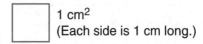
1 cm²
(Each side is 1 cm long.)

SRB
134

1.

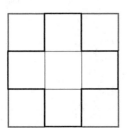

Area = _____ cm²

2.

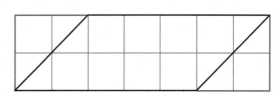

Area = _____ cm²

3.

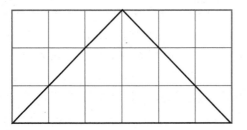

Area = _____ cm²

4.

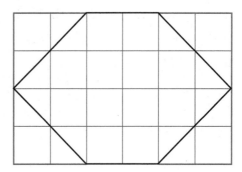

Area = _____ cm²

5.

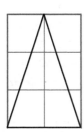

Area = _____ cm²

6.

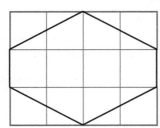

Area = _____ cm²

7.

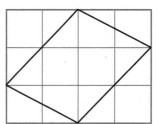

Area = _____ cm²

LESSON 8·3

Math Boxes

---

1. Use the set of numbers 8, 20, 17, 16, 5, 15, and 9 to answer the questions.

What is the

a. maximum? _____

b. minimum? _____

c. range? _____

d. median? _____

 73

---

2. Insert >, <, or = to make each number sentence true.

a. $\dfrac{1}{15}$ _____ $\dfrac{1}{6}$

b. $\dfrac{2}{3}$ _____ $\dfrac{4}{3}$

c. $\dfrac{4}{8}$ _____ $\dfrac{7}{15}$

d. $\dfrac{12}{18}$ _____ $\dfrac{4}{6}$

e. $\dfrac{7}{8}$ _____ $\dfrac{49}{50}$

 53 54

---

3. a. Use your Geometry Template to draw a regular hexagon.

b. Measure one of the angles with your protractor. Record the measure.

_____ °

 97 142 143

---

4. If you spin the spinner below 800 times, how many times would you expect it to land

on red? _____ times

on black? _____ times

on white? _____ times

on blue? _____ times

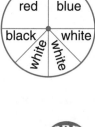

 84

---

5. Add or subtract.

a. $\dfrac{3}{10} + x = \dfrac{4}{10}$ $x =$ _____

b. $m + \dfrac{2}{3} = \dfrac{5}{6}$ $m =$ _____

c. $s - \dfrac{4}{9} = \dfrac{3}{9}$ $s =$ _____

d. $\dfrac{7}{8} - t = \dfrac{3}{4}$ $t =$ _____

 55 57

---

6. A store is giving a 50% discount on all merchandise. Find the discounted prices.

Regular price   Discounted price

$0.80   _____

$22.00   _____

$24.68   _____

$124.70   _____

 38 39

LESSON 8·4

Math Boxes

---

1. Measure the sides of the figure to the nearest centimeter. Then find its perimeter.

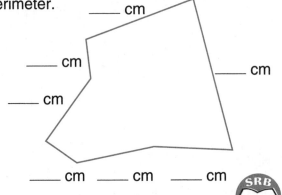

_____ cm

_____ cm

_____ cm

_____ cm

_____ cm

_____ cm _____ cm _____ cm

SRB 131

Perimeter = _____ cm

---

2. If you tossed a coin onto the grid below, about what fraction of the time would you expect it to land on a vowel?

_____

R	O	P	E
O	P	E	R
P	E	R	O
E	R	O	P

SRB 45 84

---

3. Write an equivalent fraction, decimal, or whole number.

	Decimal	Fraction
a.	0.1	_____
b.	0.20	_____
c.	_____	$\frac{4}{5}$
d.	_____	$\frac{0}{3}$

SRB 61

---

4. Which number is closest to the product of 192 and 49? Circle the best answer.

A 100

B 1,000

C 10,000

D 100,000

SRB 181

---

5. Write each number in exponential notation.

a. 100,000 = _____

b. 10 = _____

c. 10,000,000 = _____

d. 1,000,000,000 = _____

SRB 5

---

6. Shade more than $\frac{18}{100}$ but less than $\frac{3}{10}$ of the grid.

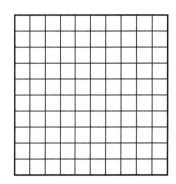

SRB 27

LESSON 8·4

What Is the Total Area of My Skin?

Follow your teacher's directions to complete this page.

1. There are _____ square inches in 1 square foot.

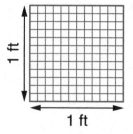

1 square foot

2. My guess is that the total
   area of my skin is about _____ square feet.

   > *Rule of Thumb:* The total area of your skin is about
   > 100 times the area of the outline of your hand.

Follow these steps to estimate the total area of your skin:

◆ Ask your partner to trace the outline of your hand on the grid on page 231.

◆ Estimate the area of the outline of your hand by counting squares on the grid.
   Record your estimate in Problem 3 below.

◆ Use the rule of thumb to estimate the total area of your skin (area of skin =
   100 * area of hand). Record your estimate in Problem 4 below.

3. I estimate that the area of the outline of my hand is about _____ square inches.

4. I estimate that the total area of my skin is about _____ square inches.

5. I estimate that the total area of my skin is about _____ square feet.

6. a. There are _____ square feet
      in 1 square yard.

   b. I estimate that the total area of my
      skin is about _____ square yards.

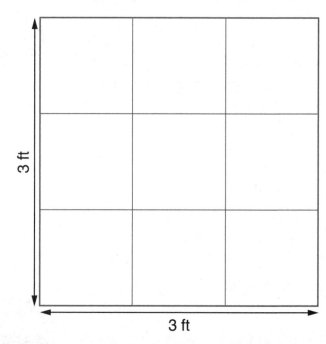

1 square yard

LESSON 8·4   What Is the Total Area of My Skin? *continued*

Ask your partner to trace the outline of your hand onto the grid below.
Keep your 4 fingers and thumb together.

Each grid square has 1-inch sides and an area of 1 square inch.

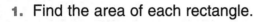

LESSON 8·5 Areas of Rectangles

Math Message

1. Find the area of each rectangle.

1 cm²

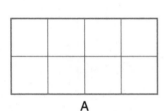

A

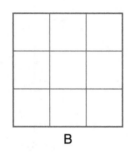

B

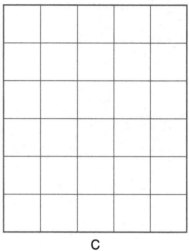

C

Area = _____ cm²   Area = _____ cm²   Area = _____ cm²

2. Fill in the table.

Rectangle	Number of squares per row	Number of rows	Total number of squares	Number model
A	4			
B				
C				

3. Write a formula for the area of a rectangle.

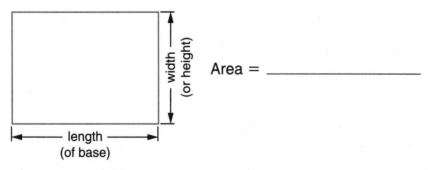

Area = _____

length (of base)

width (or height)

LESSON 8·5 **Areas of Rectangles** *continued*

4. Fill in the table at the bottom of the page.

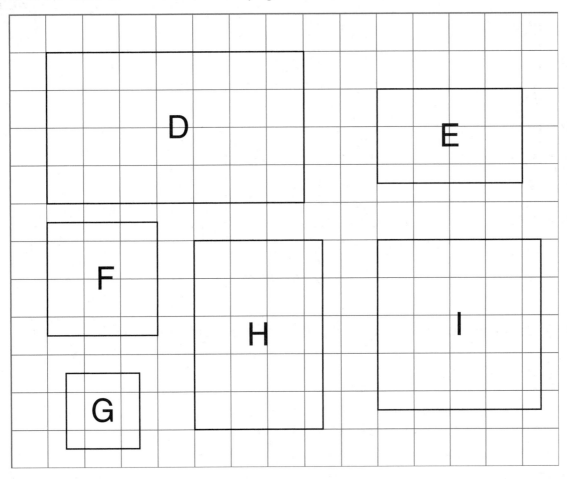

Rectangle	Area (counting squares)	Length (of base)	Width (or height)	Area (using formula)
D	_____ cm²	_____ cm	_____ cm	_____ cm²
E	_____ cm²	_____ cm	_____ cm	_____ cm²
F	_____ cm²	_____ cm	_____ cm	_____ cm²
G	_____ cm²	_____ cm	_____ cm	_____ cm²
H	_____ cm²	_____ cm	_____ cm	_____ cm²
I	_____ cm²	_____ cm	_____ cm	_____ cm²

LESSON 8·5 **Coordinate Grids**

1. Plot and label each point on the coordinate grid.

A (2,6)

B (5,5)

C (8,3)

D (4,2)

E (8,9)

F (2,10)

G (5,8)

H (1,4)

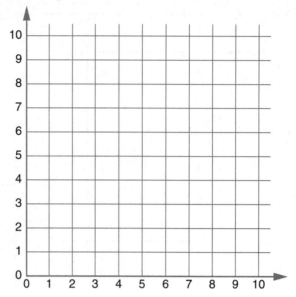

2. Write the ordered number pair for each point plotted on the coordinate grid.

I (_____ , _____)

J (_____ , _____)

K (_____ , _____)

L (_____ , _____)

M (_____ , _____)

N (_____ , _____)

O (_____ , _____)

P (_____ , _____)

Q (_____ , _____)

R (_____ , _____)

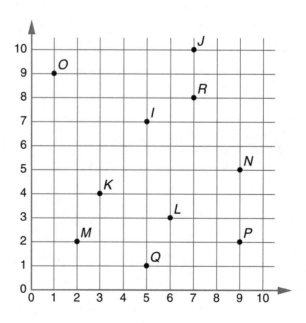

LESSON 8·5   **Math Boxes**

---

1. Write three equivalent fractions for each fraction.

a. $\frac{1}{2}$ _____, _____, _____

b. $\frac{3}{4}$ _____, _____, _____

c. $\frac{2}{3}$ _____, _____, _____

d. $\frac{5}{6}$ _____, _____, _____

SRB 49–51

---

2. Find the perimeter of this polygon.

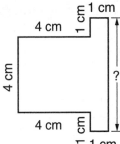

1 cm
4 cm
1 cm
4 cm
4 cm
?
4 cm
1 cm
1 cm

Number model:

_____

_____

Perimeter = _____ cm

---

3. Complete the "What's My Rule?" table, and state the rule.

Rule: _____

in	out
3.66	7.02
0.44	3.80
8.73	
	12.66

SRB 162–166

---

4. If you throw a die 60 times, about how many times would you expect to come up?

_____ times

SRB 81

---

5. Complete.

a. _____ is half as much as 44.

b. 90 is twice as much as _____.

c. _____ is 3 times as much as 40.

d. 20 is $\frac{1}{5}$ of _____.

e. _____ is 5 times as much as 34.

---

6. Divide with a paper-and-pencil algorithm.

5,682 / 4 = _____

SRB 22 23 179

LESSON 8·6 — Areas of Parallelograms

1. Cut out Parallelogram A on *Math Masters,* page 260.
   DO NOT CUT OUT THE ONE BELOW. Cut it into
   2 pieces so that it can be made into a rectangle.

1 cm²

Parallelogram A

Tape your rectangle in the space below.

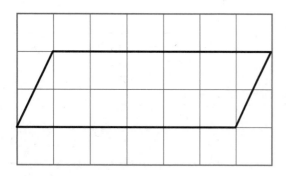

base = _____ cm

length of base = _____ cm

height = _____ cm

width (height) = _____ cm

Area of parallelogram = _____ cm²

Area of rectangle = _____ cm²

2. Do the same with Parallelogram B on *Math Masters,* page 260.

Parallelogram B

Tape your rectangle in the space below.

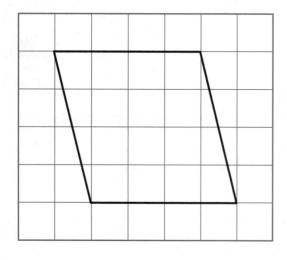

base = _____ cm

length of base = _____ cm

height = _____ cm

width (height) = _____ cm

Area of parallelogram = _____ cm²

Area of rectangle = _____ cm²

LESSON 8·6

Areas of Parallelograms *continued*

3. Do the same with Parallelogram C.

Parallelogram C

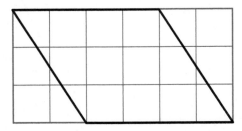

Tape your rectangle in the space below.

base = _____ cm

height = _____ cm

Area of parallelogram = _____ cm²

length of base = _____ cm

width (height) = _____ cm

Area of rectangle = _____ cm²

---

4. Do the same with Parallelogram D.

Parallelogram D

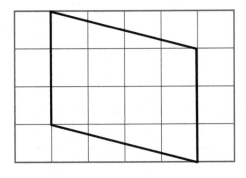

Tape your rectangle in the space below.

base = _____ cm

height = _____ cm

Area of parallelogram = _____ cm²

length of base = _____ cm

width (height) = _____ cm

Area of rectangle = _____ cm²

---

5. Write a formula for the area of a parallelogram.

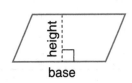

LESSON 8·6

Areas of Parallelograms *continued*

6. Draw a line segment to show the height of Parallelogram *DORA*.

   Use your ruler to measure the base and height.
   Then find the area.

   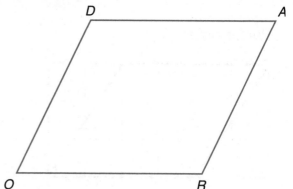

   base = _____ cm

   height = _____ cm

   Area = _____ cm²

7. Draw the following shapes on the grid below:

   a. A rectangle whose area is 12 square centimeters

   b. A parallelogram, not a rectangle, whose area is 12 square centimeters

   c. A different parallelogram whose area is also 12 square centimeters

   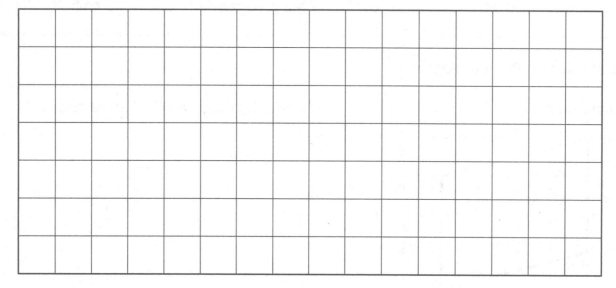

8. What is the area of:

   a. Parallelogram *ABCD*?   b. Trapezoid *EBCD*?   c. Triangle *ABE*?

   _____ cm²   _____ cm²   _____ cm²

   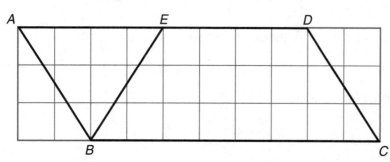

Math Boxes

1. Dimensions for actual rectangles are given. Make scale drawings of each rectangle described below.

Scale: 1 cm represents 20 meters.

a. Length of rectangle: 80 meters
Width of rectangle: 30 meters

b. Length of rectangle: 90 meters
Width of rectangle: 50 meters

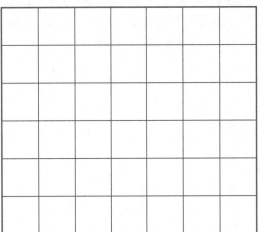

145

2. What is the area of the parallelogram?

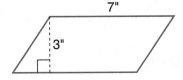

Number model: _____

Area = _____ in^2

135

3. A jar contains

8 blue blocks,

4 red blocks,

9 orange blocks, and

4 green blocks.

You put your hand in the jar and without looking pull out a block. About what fraction of the time would you expect to get a blue block?

_____

45

4. Add or subtract.

a. $\frac{3}{16} + \frac{7}{16} =$ _____

b. $\frac{2}{3} + \frac{1}{6} =$ _____

c. _____ $= \frac{9}{10} - \frac{3}{10}$

d. _____ $= \frac{3}{4} - \frac{3}{8}$

55 57

5. Multiply. Use a paper-and-pencil algorithm.

_____ $= 83 * 74$

18 19

Areas of Triangles

SRB
136

1. Cut out Triangles A and B from *Math Masters,* page 265.
    DO NOT CUT OUT THE ONE BELOW. Tape the two triangles
    together to form a parallelogram.

    1 cm²

Triangle A

Tape your parallelogram in the space below.

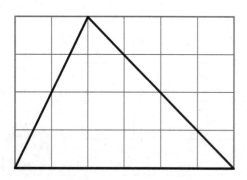

base = _____ cm

base = _____ cm

height = _____ cm

height = _____ cm

Area of triangle = _____ cm²

Area of parallelogram = _____ cm²

2. Do the same with Triangles C and D on *Math Masters,* page 265.

Triangle C

Tape your parallelogram in the space below.

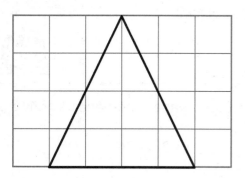

base = _____ cm

base = _____ cm

height = _____ cm

height = _____ cm

Area of triangle = _____ cm²

Area of parallelogram = _____ cm²

LESSON 8·7   **Areas of Triangles** *continued*

3. Do the same with Triangles E and F.

Triangle E

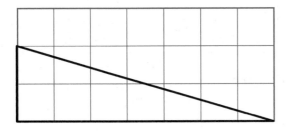

Tape your parallelogram in the space below.

base = _____ cm

base = _____ cm

height = _____ cm

height = _____ cm

Area of triangle = _____ cm^2

Area of parallelogram = _____ cm^2

4. Do the same with Triangles G and H.

Triangle G

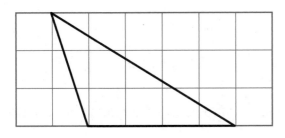

Tape your parallelogram in the space below.

base = _____ cm

base = _____ cm

height = _____ cm

height = _____ cm

Area of triangle = _____ cm^2

Area of parallelogram = _____ cm^2

5. Write a formula for the area of a triangle.

_____

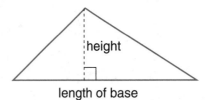

height

length of base

LESSON 8·7   **Areas of Triangles** *continued*

6. Draw a line segment to show the height of Triangle *SAM*.
   Use your ruler to measure the base and height of the
   triangle. Then find the area.

   base = _____ cm

   height = _____ cm

   Area = _____ cm²

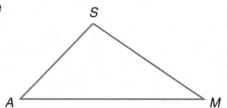

7. Draw three *different* triangles on the grid below. Each triangle must have
   an area of 3 square centimeters. One triangle should have a right angle.

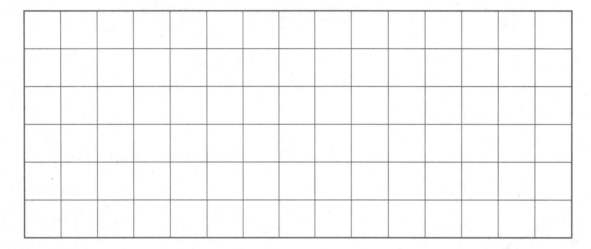

8. See the shapes below. Which has the larger area—the star or the square? Explain your answer.

   _____

   _____

   _____

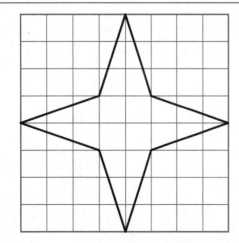
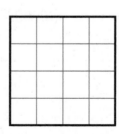

LESSON 8·7 **Fractions of Sets and Wholes**

1. Circle $\frac{1}{6}$ of the triangles. Mark Xs on $\frac{2}{3}$ of the triangles.

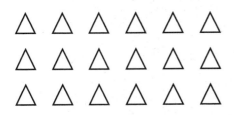

2. a. Shade $\frac{2}{5}$ of the pentagon.

   b. Shade $\frac{3}{5}$ of the pentagon.

3. There are 56 musicians in the school band: $\frac{1}{4}$ of the musicians play the flute and $\frac{1}{8}$ play the trombone.

   a. How many musicians play the flute? _____

   b. How many musicians play the trombone? _____

4. Wei had 48 bean-bag animals in her collection. She sold 18 of them to another collector. What fraction of her collection did she sell?

   _____

5. Complete.

   a. $\frac{3}{4}$ of _____ is 90.

   b. _____ of 27 is 18.

   c. $\frac{5}{6}$ of 120 is _____.

   d. $\frac{3}{10}$ of _____ is 15.

   e. _____ of 72 is 24.

   f. $\frac{5}{4}$ of 16 is _____.

6. Fill in the missing fractions on the number line.

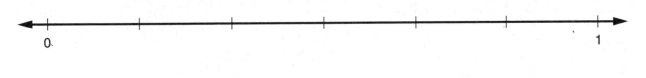

Math Boxes

1. Write three equivalent fractions for each fraction.

a. $\frac{4}{9}$ _____, _____, _____

b. $\frac{3}{8}$ _____, _____, _____

c. $\frac{2}{5}$ _____, _____, _____

d. $\frac{7}{10}$ _____, _____, _____

SRB
49–51

2. Measure the sides of the figure to the nearest centimeter to find its perimeter.

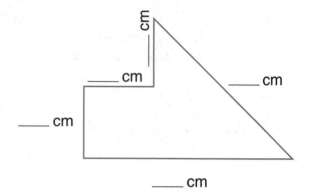

Perimeter = _____ cm

SRB
131

3. Complete the "What's My Rule?" table, and state the rule.

Rule: _____

in	out
8.69	4.09
11.03	6.43
	15.34
26.05	

SRB
162–166

4. If you throw a die 420 times, about how many times would you expect to come up? Circle the best answer.

A 70 times

B 100 times

C 50 times

D 210 times

SRB
81

5. Complete.

a. _____ is half as much as 86.

b. 48 is twice as much as _____.

c. _____ is 3 times as much as 50.

d. 40 is $\frac{1}{5}$ of _____.

e. _____ is 5 times as much as 27.

6. Divide with a paper-and-pencil algorithm.

7,653 / 6 = _____

SRB
22 23
179

244

LESSON 8·8 Comparing Country Areas

Brazil is the largest country in South America. Brazil's area is about 3,300,000 square miles. The area of the United States is about 3,500,000 square miles. So Brazil is nearly the same size as the United States.

Fill in the table below. This will help you to compare the areas of other countries in South America to Brazil's area. Round quotients in Part 4 to the nearest tenth.

Country	(1) Guess the number of times it would fit in the area of Brazil.	(2) Area	(3) Area (rounded to the nearest 10,000)	(4) Divide the rounded areas. (Brazil area ÷ country area)
Ecuador	_____	109,500 mi²	_____ mi²	3,300,000 ÷ _____ = _____
Argentina	_____	1,068,300 mi²	_____ mi²	3,300,000 ÷ _____ = _____
Paraguay	_____	157,000 mi²	_____ mi²	3,300,000 ÷ _____ = _____
Peru	_____	496,200 mi²	_____ mi²	3,300,000 ÷ _____ = _____
Uruguay	_____	68,000 mi²	_____ mi²	3,300,000 ÷ _____ = _____
Chile	_____	292,300 mi²	_____ mi²	3,300,000 ÷ _____ = _____

LESSON 8·8

Math Boxes

1. Make scale drawings of each rectangle described below.

   Scale: 1 cm represents 1.5 meters.

   a. Length of rectangle: 6 meters
      Width of rectangle: 3 meters

   b. Length of rectangle: 10.5 meters
      Width of rectangle: 4.5 meters

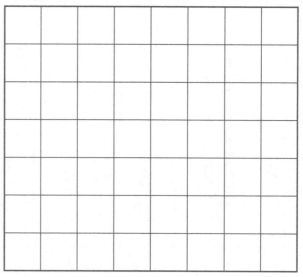

145

2. What is the area of the parallelogram?

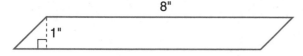

8"

1"

   Number model: _____

   Area = _____ in²

135

3. A jar contains

   27 blue blocks,

   18 red blocks,

   12 orange blocks, and

   43 green blocks.

   You put your hand in the jar and without looking pull out a block. About what fraction of the time would you expect to get a red block?

   _____

45

4. Add or subtract.

   a. $\frac{1}{12} + \frac{11}{12} =$ _____

   b. $\frac{1}{6} + \frac{2}{3} =$ _____

   c. _____ $= \frac{7}{8} - \frac{5}{8}$

   d. _____ $= \frac{5}{16} - \frac{1}{8}$

55 57

5. Multiply. Use a paper-and-pencil algorithm.

   $91 * 48 =$ _____

18 19

LESSON 8·9 **Math Boxes**

---

1. A store is giving a 50% discount on all merchandise. Find the discounted prices.

Regular price	Discounted price
$53.00	_____
$0.96	_____
$111.10	_____
$75.50	_____

SRB
38 39

2. Shade more than $\frac{70}{100}$ but less than $\frac{9}{10}$ of the grid.

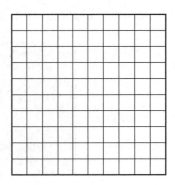

SRB
27

---

3. Multiply. Use a paper-and-pencil algorithm.

a. 482 * 6 = _____

b. 75 * 84 = _____

c. 36 * 58 = _____

SRB
18 19

---

4. Divide. Use a paper-and-pencil algorithm.

a. 853 / 7 = _____

b. 7,342 ÷ 5 = _____

c. $\frac{385}{12}$ = _____

SRB
22 23

LESSON 9·1   # Many Names for Percents

Your teacher will tell you how to fill in the percent examples.

Fill in the "100% box" for each example. Show the percent by shading the 10-by-10 grid. Then write other names for the percent next to the grid.

Example: Last season, Duncan made 62 percent of his basketball shots.

100%

all of Duncan's shots

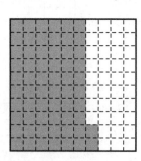

That is __62__ out of every 100.

Fraction name: $\dfrac{62}{100}$

Decimal name: __0.62__

---

1. Percent Example: _____

_____

_____

100%

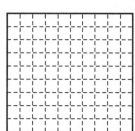

That is _____ out of every 100.

Fraction name: $\dfrac{\boxed{}}{100}$

Decimal name: _____

---

2. Percent Example: _____

_____

_____

100%

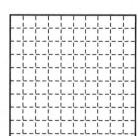

That is _____ out of every 100.

Fraction name: $\dfrac{\boxed{}}{100}$

Decimal name: _____

LESSON 9·1   **Many Names for Percents** *continued*

Fill in the "100% box" for each example. (Problem 3 is done for you.)
Show the percent by shading the 10-by-10 grid. Then write other names
for the percent next to the grid.

3. Example: 12% of the students in Marshall School are left-handed.

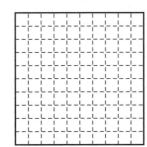

100%

*all students at
Marshall School*

That is _____ out of every 100.

Fraction name: $\dfrac{\boxed{}}{100}$

Decimal name: _____

---

4. Example: Sarah spelled 80% of the words correctly on her last test.

100%

That is _____ out of every 100.

Fraction name: $\dfrac{\boxed{}}{100}$

Decimal name: _____

---

5. Example: Cats sleep about 58% of the time.

100%

That is _____ out of every 100.

Fraction name: $\dfrac{\boxed{}}{100}$

Decimal name: _____

LESSON 9·1   **Many Names for Percents** *continued*

Fill in the "100% box" for each example. Show the percent by shading
the 10-by-10 grid. Write other names for the percent next to the grid.
Then answer the question.

6. Example: *Sale—40% Off*
                  Everything Must Go!

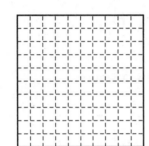

100%

That is _____ out of every 100.

Fraction name: $\dfrac{\boxed{}}{100}$

Decimal name: _____

What would Mr. Thompson pay for a bicycle that had been selling for $300? _____

---

7. Example: A carpet store ran a TV commercial that said:
          "Pay 20% when you order. Take 1 full year to pay the rest."

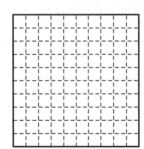

100%

That is _____ out of every 100.

Fraction name: $\dfrac{\boxed{}}{100}$

Decimal name: _____

Mrs. Shields wants to order a $1,200 carpet. How much must she pay

when she orders it? _____

LESSON 9·1 **Math Boxes**

---

1. In which situation below do you need to know the area? Choose the best answer.

   ⬭ finding the distance around a pool

   ⬭ buying a wallpaper border for your bedroom

   ⬭ carpeting the living room

   ⬭ fencing a yard

   SRB
   131 133

---

2. Complete.

   Rule:

   _____

in	out
$\frac{2}{5}$	$\frac{3}{5}$
$\frac{1}{10}$	$\frac{3}{10}$
$\frac{4}{5}$	
	$\frac{9}{10}$
	$\frac{1}{2}$

   SRB
   162–166

---

3. Multiply. Use a paper-and-pencil algorithm.

   _____ = 58 * 76

   SRB
   18   19

---

4. Find the approximate latitude and longitude of these Region 2 cities.

   a. Dublin, Ireland   latitude _____° _____

   longitude _____° _____

   b. Rome, Italy   latitude _____° _____

   longitude _____° _____

   SRB
   272 273

---

5. Angle RST is an _____ (acute or obtuse) angle.

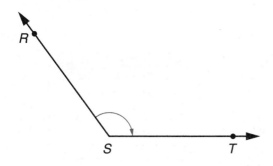

The measure of ∠RST is _____°.

SRB
141–143

---

6. Use a straightedge to draw the line of symmetry.

SRB
109

LESSON 9·2 **"Percent-of" Number Stories**

Alfred, Nadine, Kyla, and Jackson each took the same math test. There were 20 problems on the test.

100%

20-problem test

1. Alfred missed $\frac{1}{2}$ of the problems. He missed 0.50 of the problems. That is 50% of the problems.

   How many problems did he miss? _____ problems

   $\frac{1}{2}$ of 20 = _____

   50% of 20 = _____

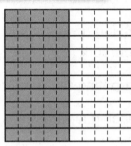

$\frac{1}{2}$, or 50% is shaded.

2. Nadine missed $\frac{1}{4}$ of the problems. She missed 0.25 of the problems. That is 25% of the problems.

   How many problems did she miss? _____ problems

   $\frac{1}{4}$ of 20 = _____

   25% of 20 = _____

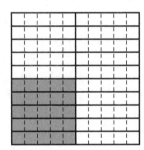

$\frac{1}{4}$, or 25% is shaded.

3. Kyla missed $\frac{1}{10}$ of the problems. She missed 0.10 of the problems. That is 10% of the problems.

   How many problems did she miss? _____ problems

   $\frac{1}{10}$ of 20 = _____

   10% of 20 = _____

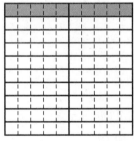

$\frac{1}{10}$, or 10% is shaded.

4. Jackson missed $\frac{1}{5}$ of the problems. He missed 0.20 of the problems. That is 20% of the problems.

   How many problems did he miss? _____ problems

   $\frac{1}{5}$ of 20 = _____

   20% of 20 = _____

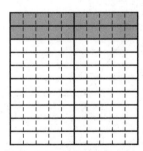

$\frac{1}{5}$, or 20% is shaded.

Date _____   Time _____

Fractions, Decimals, and Percents

Fill in the missing numbers. Problem 1 has been done for you.

100%

large square

1. Ways of showing $\frac{3}{4}$:

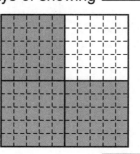

$\frac{3}{4}$ is shaded. $\frac{75}{100}$

0.**75**   **75** %

2. Ways of showing _____:

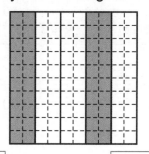

$\frac{\boxed{}}{5}$ is shaded. $\frac{\boxed{}}{100}$

0._____   _____%

3. Ways of showing _____:

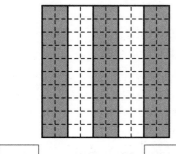

$\frac{\boxed{}}{5}$ is shaded. $\frac{\boxed{}}{100}$

0._____   _____%

4. Ways of showing _____:

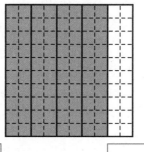

$\frac{\boxed{}}{5}$ is shaded. $\frac{\boxed{}}{100}$

0._____   _____%

5. Ways of showing _____:

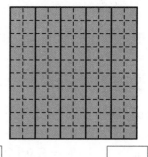

$\frac{\boxed{}}{5}$ is shaded. $\frac{\boxed{}}{100}$

_____   _____%

Shade the grid. Then fill in the missing numbers.

6. Ways of showing $\frac{3}{10}$:

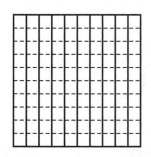

$\frac{\boxed{}}{100}$ is shaded.

0._____   _____%

7. Ways of showing $\frac{7}{10}$:

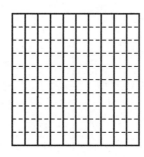

$\frac{\boxed{}}{100}$ is shaded.

0._____   _____%

8. Ways of showing $\frac{9}{10}$:

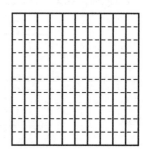

$\frac{\boxed{}}{100}$ is shaded.

0._____   _____%

LESSON 9·2   **Math Boxes**

---

1. Complete the table with equivalent names.

Fraction	Decimal	Percent
$\frac{1}{5}$		
	0.80	
$\frac{3}{10}$		
		90%

SRB
61 62

---

2. About 4.02% of the words on the Internet are *the,* and about 1.68% of the words are *and.* About what percent of all words on the Internet are either *the* or *and?* Choose the best answer.

⬭ 5.71%

⬭ 5.7%

⬭ 570%

⬭ 57%

SRB
34–37

---

3. Complete.

a. 3 yd 2 ft = _____ ft

b. 6 yd 1 ft = _____ ft

c. _____ in. = 2 yd

d. _____ ft = 5 yd 2 ft

e. 25 ft = _____ yd _____ ft

f. _____ ft _____ in. = 30 in.

SRB
129

---

4. Zena earned $12. She spent $8.

a. What fraction of her earnings did she spend? _____

b. What fraction did she have left? _____

c. The amount she spent is how many times as much as the amount she saved?

_____

SRB
44

---

5. Find the area and perimeter of the rectangle. Include the correct units.

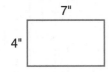

Area = _____

Perimeter = _____

SRB
131 133

---

6. What temperature is it?

_____°F

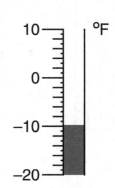

SRB
139

LESSON 9·3 **Math Boxes**

---

1. Write A or P to tell whether you would need to find the area or the perimeter in each situation.

a. buying a garden fence _____

b. finding the square footage of your bedroom _____

c. buying wallpaper for the kitchen _____

SRB 131 133

---

2. Complete.

Rule:

_____

in	out
$\frac{7}{9}$	$\frac{4}{9}$
	$\frac{1}{9}$
$\frac{9}{18}$	
$\frac{6}{9}$	$\frac{1}{3}$

SRB 162–166

---

3. Multiply. Use a paper-and-pencil algorithm.

_____ = 64 * 23

SRB 18 19

---

4. Find the approximate latitude and longitude of these Region 4 cities.

a. Calcutta, India latitude _____ ° _____

longitude _____ ° _____

b. Seoul, Korea latitude _____ ° _____

longitude _____ ° _____

SRB 272 273

---

5. Angle *RLA* is an _____ (acute or obtuse) angle.

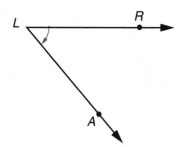

The measure of ∠ *RLA* is _____ °.

SRB 141–143

---

6. Use a straightedge to draw the line of symmetry.

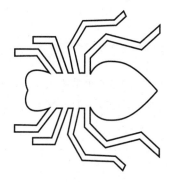

SRB 109

LESSON 9·4   Discount Number Stories

1. A store is offering a **discount** of 10% on all items. This means that you save $\frac{1}{10}$ of the **regular price.** Find the sale price of each item below. The **sale price** is the amount you pay after subtracting the discount from the regular price.

Item	Regular Price	Discount (10% of regular price)	Sale Price (Subtract: regular price – discount)
CD player	$140	$14	
Giant screen TV	$1,200		
Radio	$80		
DVD player		$3	

2. Use a drawing and number models to explain how you found the discount and sale price for the radio.

    _____

    _____

3. An airline offers a 25% discount on the regular airfare for tickets purchased at least 1 month in advance. Find the sale price of each ticket below.

Regular Airfare	Discount (25% of regular airfare)	Sale Price (Subtract: regular airfare – discount)
$400	$100	
$240		
	$75	

4. Use a drawing and number models to explain how you found the regular airfare when you knew $75 was 25% of the regular airfare.

    _____

    _____

LESSON 9·4   **Discount Number Stories** *continued*

5. The regular price of a swing set is $400. Mrs. Lefevre received a
   30% discount because she ordered it during the Big Spring Sale.

   a. How much did she save? _____

   b. How much did she pay? _____

   c. Explain how you solved the problem.

   _____

   _____

   _____

Try This

6. You can pay for a refrigerator by making 12 payments of $50 each.
   Or you can save 25% if you pay for it all at once.

   a. How much will the refrigerator cost if you pay for it all at once? _____

   b. Explain how you solved the problem.

   _____

   _____

   _____

7. Write your own discount number story. Ask a partner to solve it.

   _____

   _____

   _____

   _____

LESSON 9·4 Cellular Telephone Use

1. Use the "Subscriptions per 100 People" data in the Cellular Telephone Use table at the bottom of *Student Reference Book*, page 302 to complete the bar graph. Round each decimal to the nearest whole number.

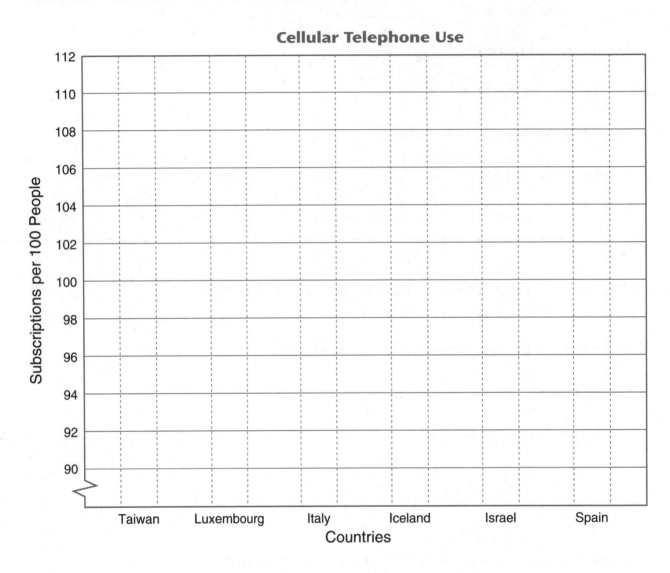

Cellular Telephone Use

2. Write a question that can be answered by looking at the data displayed in the bar graph. Answer the question.

_____

_____

_____

_____

LESSON 9·4

Math Boxes

1. Complete the table with equivalent names.

Fraction	Decimal	Percent
$\frac{3}{4}$		
	0.6	
$\frac{1}{10}$		
		50%

SRB
61 62

2. About 1.96% of the words on the Internet are *a*, and about 0.81% of the words are *it*. What percent of all words on the Internet are either *a* or *it*? Show your work.

_____

SRB
34–37

3. 2 ft 7 in. = __ in. Choose the best answer.

◯ 14

◯ 24

◯ 27

◯ 31

SRB
129

4. Three girls cut a pizza into 12 equal slices and plan to share the pizza equally.

a. What fraction of the pizza should each girl get? _____

b. How many slices should each girl get? _____ slices

c. Suppose 2 more girls arrive. How many slices should each of the 5 girls get?

_____ slices

SRB
44

5. Find the area and perimeter of the rectangle. Include the correct units.

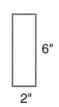

6"

2"

Area = _____

Perimeter = _____

SRB
131 133

6. What temperature is it?

_____ °C

SRB
139

LESSON 9·5   **Math Boxes**

---

1. Calculate.

   a. 10% of 70 = _____

   b. 5% of 60 = _____

   c. 25% of _____ = 7

   d. _____% of 48 = 24

   e. _____% of 25 = 20
   SRB
   38 39

2. Insert parentheses to make each number sentence true.

   a. 6 + 2 * 4 = 32

   b. 5 + 7 * 3 = 36

   c. 1 + 8 * 8 + 2 = 90

   d. 1 + 7 * 8 + 2 = 80
   SRB
   150

---

3. Complete the table with equivalent names.

Fraction	Decimal	Percent
$\frac{5}{10}$		
	0.20	
		70%
	0.4	

SRB
61 62

4. Divide. Use a paper-and-pencil algorithm.

   897 ÷ 6 = _____

SRB
22 23

---

5. What is the height of the parallelogram? Include the correct unit.

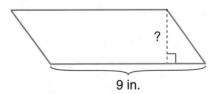

   9 in.

   Area = 27 in²

   Number model: _____

   Height: _____
   SRB
   135

6. Draw the mirror image of the figure shown on the left of the vertical line.

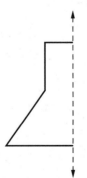

SRB
106 109

LESSON 9·5 — Finding Unknown Angle Measures

Without using a protractor, find the measure of the unknown angle. Write an equation to show how you solved the problem. Use a variable to represent the unknown angle measure.

1.

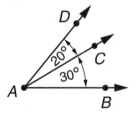

Measure of ∠DAB = _____°

Equation: _____

2.

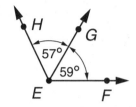

Measure of ∠HEF = _____°

Equation: _____

3.

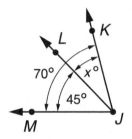

Measure of ∠KJL = _____°

Equation: _____

4.

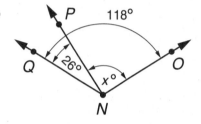

Measure of ∠PNO = _____°

Equation: _____

5.

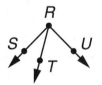

∠SRU is a right angle.

Measure of ∠TRU = 60°

Measure of ∠TRS = _____°

Equation: _____

6.

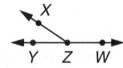

∠YZW is a straight angle.

Measure of ∠XZY = 33°

Measure of ∠XZW = _____°

Equation: _____

LESSON 9·5 | **Finding Unknown Angle Measures** *continued*

Find the value of *x*. Write an open sentence to show how you solved the problem.

7.

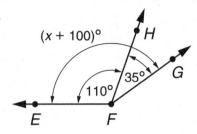

$x =$ _____

Equation: _____

Measure of $\angle EFG =$ _____ °

8.

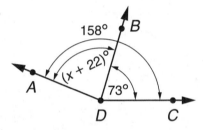

$x =$ _____

Equation: _____

Measure of $\angle ADB =$ _____ °

Try This

9. Angela used her protractor to measure $\angle KJL$ and $\angle NJM$. She found the two angles had the same measure. Angela said, "Without measuring, I also know $\angle KJM$ has the same measure as $\angle NJL$."

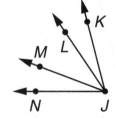

Explain how Angela knows her statement is true.

_____

_____

_____

_____

LESSON 9·6 # Trivia Survey Results

1. The chart below will show the results of the trivia survey for the whole class. Wait for your teacher to explain how to fill in the chart.

Class Results for the Trivia Survey

Question	Yes	No	Total	$\frac{Yes}{Total}$	% Yes
1. Is Monday your favorite day?					
2. Have you gone to the movies in the last month?					
3. Did you eat breakfast today?					
4. Do you keep a map in your car?					
5. Did you eat at a fast-food restaurant yesterday?					
6. Did you read a book during the last month?					
7. Are you more than 1 meter tall?					
8. Do you like liver?					

2. On the basis of the survey results, is it more likely that a person will

a. read a book or go to a movie?

_____

b. eat breakfast or eat at a fast-food restaurant?

_____

c. like liver or like Mondays?

_____

LESSON 9·6

Planting a Vegetable Garden

SRB
178A
178B

Kim and Carl decide to plant a vegetable garden. Solve the following problems about their garden.

1. Kim and Carl plan to make their garden in the shape of a rectangle that is 8 feet long. They have 36 feet of fencing. How wide should their garden be if they want to use all the fencing?

_____   Answer: _____ feet wide
(number model(s) with unknown)

_____
(summary number model(s))

2. Pepper plants are on sale. The original price for 6 plants was $14.95. The sale price for 6 plants is $9.72. If they purchase 12 plants on sale, how much will they spend?

_____   Answer: _____
(number model(s) with unknown)

_____
(summary number model(s))

3. Some small vegetable seeds come attached to a 15-foot tape instead of in a seed packet. Kim purchased lettuce, carrot, and radish seed tapes, each costing $4.95. What is the cost per foot for a seed tape?

_____   Answer: _____
(number model(s) with unknown)

_____
(summary number model(s))

4. Carl told Kim that a radish takes about 600 hours to grow after the seed is planted. About how long, in weeks and days, does Kim have to wait to eat a radish after she plants the seeds?

_____
(number model(s) with unknown)

Answer: _____ weeks _____ days

_____
(summary number model(s))

LESSON 9·6   Planting a Vegetable Garden *continued*

SRB
178A
178B

5. The tomatoes take about 77 days to grow. The leaf lettuce takes 45 days to grow. How many hours longer does it take the tomatoes to grow than the leaf lettuce?

_____   Answer: _____ hours
(number model(s) with unknown)

_____
(summary number model(s))

6. Carl and Kim's garden was a great success. Kim and Carl picked 132 tomatoes from their garden. They put aside a third of the tomatoes to make spaghetti sauce. If each batch of spaghetti sauce uses 10 tomatoes, how many full batches of sauce will they be able to make?

_____   Answer: _____ batches
(number model(s) with unknown)

_____
(summary number model(s))

7. They sold most of the carrots at farmer's markets. They tied the carrots into bundles of 6. Then they placed 8 bundles into each basket. If they sold 15 baskets of carrots during the season, how many carrots did they sell in all?

_____   Answer: _____ carrots
(number model(s) with unknown)

_____
(summary number model(s))

Try This

8. They picked about 23 pounds of green beans over the summer. The family ate about 12 pounds of green beans. Their mother froze the rest in plastic bags weighing about 22 ounces each. About how many bags of green beans did their mother freeze? (*Hint:* There are 16 ounces in a pound.)

_____
(number model(s) with unknown)

Answer: _____ bags

_____
(summary number model(s))

261B

LESSON 9·6   **Math Boxes**

1. If you threw a 6-sided die 54 times, about how many times would you expect it to land on a number less than 3? Choose the best answer.

   ⬭   9 times

   ⬭   12 times

   ⬭   18 times

   ⬭   36 times

   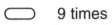 SRB 81

2. Name a percent value

   a. greater than $\frac{1}{4}$ and less than $\frac{2}{3}$.

   _____

   b. less than $\frac{4}{5}$ and greater than $\frac{5}{8}$.

   _____

    SRB 61 62

3. Store X is selling bathing suits at 20% off the regular price of $35. Store Y is selling the same suits for $\frac{1}{4}$ off the regular price of $32. Which store is offering the better buy?

   _____

   Show how you solved the problem.

    SRB 38 39 59

4. If 1 inch on a map represents 200 miles, then

   a. 5 inches represent _____ miles.

   b. 8 inches represent _____ miles.

   c. _____ inches represent 800 miles.

   d. $3\frac{1}{4}$ inches represent _____ miles.

   e. _____ inches represent 350 miles.

    SRB 145

5. What is the area of the triangle? Include the correct unit.

   6"
   8"

   Number model: _____

   Area = _____   SRB 136

6. a. Which is warmer, −15°C or −3°C?

   _____

   How many degrees warmer?

   _____

   b. Which is colder, −15°C or −20°C?

   _____

   How many degrees colder?

   _____   SRB 60 139

LESSON 9·7

Math Boxes

---

1. Calculate.

   a. 10% of 50 = _____

   b. 5% of 80 = _____

   c. 20% of _____ = 8

   d. _____% of 16 = 12

   e. _____% of 24 = 6

38   39

---

2. Insert parentheses to make each number sentence true.

   a. 63 / 21 − 12 = 7

   b. 34 − 8 + 4 = 22

   c. 3 * 5 + 6 > 3 * 10

   d. 6 * 7 + 1 > (80 / 2) + 5

150

---

3. Complete the table with equivalent names.

Fraction	Decimal	Percent
		63%
	1.00	
$\frac{3}{5}$		
		80%

61   62

---

4. Divide. Use a paper-and-pencil algorithm.

420 ÷ 16 = _____

22 23

---

5. What is the area of the parallelogram? Include the correct unit.

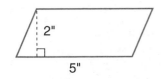

2"

5"

Number model: _____

Area = _____

135

---

6. Draw the mirror image of the figure shown on the left of the vertical line.

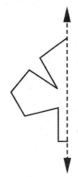

106 109

263

LESSON 9·7 Color-Coded Population Maps

301

1. List the countries in Region 4 from *least to greatest* according to the **percent of population, ages 0–14.** Take one copy of *Math Masters,* page 293 and write a title for your first map. Color these countries using the color code shown below.

Rank	Country	Percent of Population Ages 0–14	Color Code
1	*Japan*	*15%*	blue
2			blue
3			blue
4			green
5			green
6			green
7			green
8			red
9			red
10	*Bangladesh*	*34%*	red

2. List the countries in Region 4 from *least to greatest* according to the **percent of population that is rural.** Take another copy of *Math Masters,* page 293 and write a title for your second map. Color these countries using the color code shown below.

Rank	Country	Percent of Rural Population	Color Code
1			blue
2			blue
3			blue
4			green
5			green
6			green
7			green
8			red
9			red
10			red

LESSON 9·7   **Color-Coded Population Maps** *continued*

3. What do the map colors mean?

    a. blue _____

    b. green _____

    c. red _____

4. **a.** Which countries are colored blue on both maps?

    _____

    b. What does this tell you about these countries?

    _____

    _____

5. **a.** Which countries are colored red on both maps?

    _____

    b. What does this tell you about these countries?

    _____

    _____

6. **a.** Based on the data tables and maps, do you think there is a connection
between the percent of young people who live in a particular country
and the percent of people who live in rural areas? _____

    Explain your answer.

    _____

    _____

    _____

    _____

    _____

    b. How does your conclusion compare with the prediction that you made
at the beginning of the lesson?

    _____

    _____

Date _____   Time _____

1. For this spinner, about what fraction of the spins would you expect to land on blue?

_____ of the spins

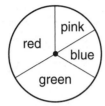

2. Design a spinner so that the probability of landing on red is $\frac{3}{8}$, on blue is $\frac{1}{4}$, on yellow is $\frac{1}{8}$, and on green is $\frac{1}{4}$.

3. If you throw a 6-sided die 84 times, how many times would you expect a 4 to come up?

_____ times

Circle the word that best describes your chances of rolling a 4.

certain

likely

very unlikely

4. If you tossed a coin onto the grid below, what fraction of the tosses would you expect it to land on M?

_____ of the tosses

M	A	P
P	M	A
A	P	M

5. If you throw a 6-sided die 96 times, how many times would you expect it to land on a factor of 24?

_____ times

Explain. _____

_____

_____

_____

6. You flip a coin 9 times, and each time the coin lands on heads. If you flip the same coin one more time, is it more likely to come up heads than tails or just as likely to come up heads as tails? Explain.

_____

_____

_____

_____

LESSON 9·8

Math Boxes

1. **a.** If you threw a 6-sided die 48 times, about how many times would you expect it to land on a number greater than or equal to 4?

_____ times

b. If you threw a 6-sided die 54 times, about how many times would you expect it to land on a number greater than 4?

_____ times

SRB 81

2. Name a percent value

a. greater than $\frac{1}{5}$ and less than $\frac{1}{2}$.

_____

b. less than $\frac{3}{4}$ and greater than $\frac{3}{5}$.

_____

SRB 61 62

3. Homer's is selling roller blades at 25% off the regular price of $52.00. Martin's is selling them for $\frac{1}{3}$ off the regular price of $60. Which store is offering the better buy?

_____

Show how you solved the problem.

SRB 38 39 59

4. If 1 centimeter on a map represents 300 kilometers, then 2.5 centimeters represents ____ kilometers. Choose the best answer.

⬭ 600

⬭ 650

⬭ 350

⬭ 750

SRB 145

5. What is the area of the triangle? Include the correct unit.

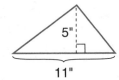

5"

11"

Number model: _____

Area = _____

SRB 136

6. **a.** Which is warmer, −7°C or −3.5°C?

_____

How many degrees warmer?

_____

b. Which is colder, −18°C or −9.6°C?

_____

How many degrees colder?

_____

SRB 60 139

LESSON 9·8 Multiplying Decimals

Math Message

Toni has 8 blocks. Each block is 1.2 centimeters high.
If she stacks the blocks, what will be the height of the stack?   _____ cm

---

1. Devon measured the length of the room by pacing it off.
   The length of his pace was 2.3 feet. He counted 14 paces.
   How long is the room?   _____ ft

2. Spiral notebooks are on sale for $0.35 each.
   How much will 25 spiral notebooks cost? $_____

3. Find the area of each rectangle below. Include the correct unit.
   Write a number model to show how you found the answer.

   a. 1.5 cm [rectangle]
       30 cm

   Number model: _____   Area = _____

   b. 6 in. [rectangle]
       15.4 in.

   Number model: _____   Area = _____

4. For each problem below, the multiplication has been done correctly, but the decimal point
   is missing in the answer. Write a number model to show how you estimated the answer.
   Then correctly place the decimal point in the answer.

   a. $23 * 7.3 =$ 1 6 7 9   b. $6.91 * 82 =$ 5 6 6 6 2

       Number model: Number model:

       _____   _____

   c. $5,203 * 12.6 =$ 6 5 5 5 7 8   d. $0.38 * 51 =$ 1 9 3 8

       Number model: Number model:

       _____   _____

LESSON 9·8 Multiplying Decimals *continued*

Write a number model to estimate each product. Then multiply the factors as though they were whole numbers. Use your estimate to help you place the decimal in the answer.

5. 2.7 * 45 = _____

Number model: _____

6. 8 * 5.7 = _____

Number model: _____

7. 5.08 * 27 = _____

Number model: _____

8. 42 * 0.97 = _____

Number model: _____

Try This

9. 22 * 0.32 = _____

Number model:

_____

10. 0.02 * 333 = _____

Number model:

_____

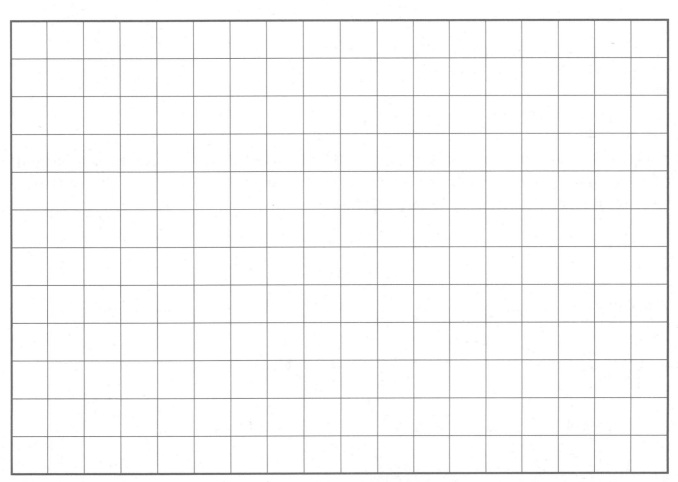

LESSON 9·9 Dividing Decimals

1. Janine is building a bookshelf. She has a board that is 3.75 meters long. She wants to cut it into 5 pieces of equal length. What will be the length of each piece?

   _____ meters

2. Three sisters set up a lemonade stand. On Wednesday they made $8.46. If they shared the money equally, how much did each girl get?

   $_____

3. Alex and his three friends went out to lunch. The total bill, including tax and tip, was $42.52. They decided that each would pay the same amount. How much did each person pay?

   $_____

4. Victor divides a 98.4 cm piece of string into 3 equal pieces. What is the length of each piece?

   _____ centimeters

For each problem below, the division has been done correctly, but the decimal point is missing in the answer. Write a number model to show how to estimate the answer. Use your estimate to correctly place the decimal point in the answer.

5. $$\begin{array}{r} 1\,4\,6 \\ 3\overline{)4\,3.8} \end{array}$$

   Number model: _____

6. $$\begin{array}{r} 4\,9\,8\,2 \\ 6\overline{)2\,9\,8.9\,2} \end{array}$$

   Number model: _____

7. $$\begin{array}{r} 1\,6\,1\,5 \\ 4\overline{)6.4\,6} \end{array}$$

   Number model: _____

8. $$\begin{array}{r} 8\,7 \\ 5\overline{)4.3\,5} \end{array}$$

   Number model: _____

LESSON 9·9 · Dividing Decimals *continued*

Write a number model to estimate each quotient. Then divide the numbers as though they were whole numbers. Use the estimate to help you place the decimal point in the answer.

9. _____ = 9.44 / 4

   Number model: _____

10. 89.6 / 4 = _____

    Number model: _____

11. 46.8 ÷ 12 = _____

    Number model: _____

12. 253.8 / 6 = _____

    Number model: _____

Try This

13. 2.96 / 8 = _____

    Number model:

    _____

14. _____ = 3.65 ÷ 5

    Number model:

    _____

LESSON 9·9   Math Boxes

---

1. Calculate.

  a. 10% of 90 = _____

  b. 5% of 140 = _____

  c. _____% of 30 = 24

  d. _____% of 48 = 36

  e. 20% of _____ = 9

---

2. Insert parentheses to make each number sentence true.

  a. 4 * 6 + 3 > 3 * 10

  b. 57 − 24 + 15 = 18

  c. 40 * 30 + 60 > 100 * 20

  d. 56 / 7 − 3 = 14

---

3. Complete the table with equivalent names.

Fraction	Decimal	Percent
		70%
	0.75	
$\frac{3}{5}$		
		72%

---

4. Divide. Use a paper-and-pencil algorithm.

268 ÷ 12 = _____

---

5. What is the area of the parallelogram? Include the correct unit.

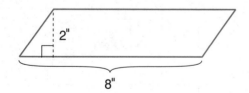

2"

8"

Number model: _____

Area = _____

---

6. Study the figure. Draw the other half along the vertical line of symmetry.

LESSON 9·10

Math Boxes

1. Use a straightedge to draw the line of symmetry.

SRB
109

2. What temperature is it?

_____ °F

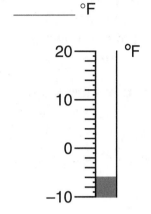

SRB
139

3. Draw the mirror image of the figure shown on the left of the vertical line.

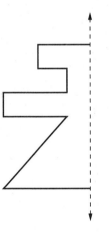

SRB
106 109

4. **a.** Which is warmer, −9.4°C or −11.2°C?

_____

b. How many degrees warmer?

_____

c. Which is colder, −19.3°C or −12.8°C?

_____

d. How many degrees colder?

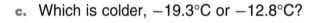

SRB
60 139

5. Name four numbers greater than −8 and less than −5.

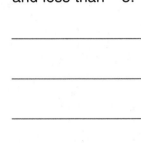

SRB
60

6. Name four numbers less than 2 and greater than −1.

_____

_____

_____

_____

SRB
60

LESSON 10·1   **Basic Use of a Transparent Mirror**

A **transparent mirror** is shown at the right.

Notice that the mirror has a **recessed** drawing edge, along which lines are drawn. Some transparent mirrors have a drawing edge both on the top and on the bottom.

Place your transparent mirror on this page so that its drawing edge lies along line *MK* below. Then look through the transparent mirror to read the "backward" message.

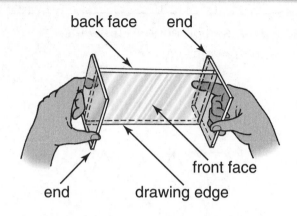

back face   end

end   drawing edge

front face

M ←————————————————————————————→ *K*

◆ Experiment and have fun!

◆ Use a sharp pencil when tracing along the drawing edge.

◆ Use your transparent mirror on flat surfaces like your desk or a tabletop. In this position, the drawing edge will be facing you.

◆ Always look into the front of the transparent mirror.

mirror:

message. Here are a few things to remember when using your transparent

If you have followed the directions correctly, you are now able to read this

274

LESSON 10·1   Math Boxes

1. Use your Geometry Template to draw the image of the figure that is shown on the right of the line of reflection.

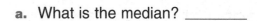
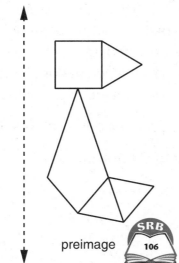

image   preimage   **SRB** 106

2. The following totals came up when Tina threw two dice:

4, 5, 9, 6, 12, 12, 2, 5, 6, 12, 3

   a. What is the median? _____

   b. Mode? _____

   c. Maximum? _____

   d. Minimum? _____

   e. Range? _____

SRB 73

3. Order the fractions from least to greatest.

   a. $\dfrac{8}{7}, \dfrac{5}{7}, \dfrac{3}{7}, \dfrac{1}{7}, \dfrac{4}{7}$ _____

   b. $\dfrac{2}{4}, \dfrac{2}{3}, \dfrac{2}{5}, \dfrac{2}{9}, \dfrac{2}{12}$ _____

   c. $\dfrac{3}{5}, \dfrac{9}{10}, \dfrac{4}{9}, \dfrac{1}{2}, \dfrac{2}{7}$ _____

SRB 53 54

4. Insert the decimal point in each product.

   a. $8 * 9.6 =$ 7 6 8

   b. 9 4 2 4 $= 19 * 4.96$

   c. 7 4 2 $= 14 * 0.53$

   d. $83 * 0.02 =$ 1 6 6

SRB 18 19

5. Miki took 40 shots in a basketball game. She missed 30% of the shots that she took.

   a. What fraction of the shots did she miss?

   _____

   b. How many shots did she miss?

   _____ shots

SRB 38 39 62

6. What are the perimeter and area of the rectangle? Include the correct units.

3 ft

18 ft

Perimeter = _____

Area = _____

SRB 131 134

LESSON 10·1 Using Coins to Add Fractions

In the United States, each coin is worth a fraction of a dollar. For example, the penny is worth $\frac{1}{100}$ of a dollar. The values of some coins can be expressed as a fraction of a dollar in different ways. Two ways to express the value of a quarter are as $\frac{1}{4}$ of a dollar and as $\frac{25}{100}$ of a dollar.

When a coin's value is written as a *unit fraction* (a fraction with a numerator of 1), the denominator tells you how many of those coins are needed to make a dollar. When a coin's value is written as a fraction with a denominator of 100, the numerator tells you how many pennies are needed to equal the value of the coin.

Find the missing numbers to show the value of each coin as a fraction of a dollar.

Coin	Unit fraction	Denominator of 100
Penny:	$\frac{1}{\boxed{}}$	$\frac{\boxed{}}{100}$
Nickel:	$\frac{1}{\boxed{}}$	$\frac{\boxed{}}{100}$
Dime:	$\frac{1}{\boxed{}}$	$\frac{\boxed{}}{100}$
Quarter:	$\frac{1}{\boxed{}}$	$\frac{\boxed{}}{100}$
Half Dollar:	$\frac{1}{\boxed{}}$	$\frac{\boxed{}}{100}$

For Problems 1–3 on the next page, rewrite the equation to show value of each coin as a unit fraction. Then rename the fractions as hundredths and add to find the total value of the coins.

Example:

$(P) + (Q) + (N) + (N) + (HD) = ?$

Unit fractions: $\frac{1}{100} + \frac{1}{4} + \frac{1}{20} + \frac{1}{20} + \frac{1}{2} = ?$

Hundredths: $\frac{1}{100} + \frac{25}{100} + \frac{5}{100} + \frac{5}{100} + \frac{50}{100} = \frac{86}{100}$, or 86 cents

LESSON 10·1   **Using Coins to Add Fractions** *continued*

1. $\text{N} + \text{D} + \text{P} + \text{D} + \text{Q} = ?$

   Unit fractions: _____

   Hundredths: _____

2. $\text{Q} + \text{Q} + \text{D} + \text{D} + \text{P} = ?$

   Unit fractions: _____

   Hundredths: _____

3. $\text{P} + \text{N} + \text{D} + \text{Q} + \text{HD} = ?$

   Unit fractions: _____

   Hundredths: _____

For Problems 4–6, the fractions represent coin values. Rename them as hundredths and then add to find the total value.

4. $\frac{1}{2} + \frac{1}{20} + \frac{5}{100} + \frac{10}{100} + \frac{1}{100} + \frac{1}{100}$

   Hundredths: _____

5. $\frac{1}{2} + \frac{1}{4} + \frac{1}{10} + \frac{1}{20} + \frac{1}{100}$

   Hundredths: _____

6. $\frac{1}{4} + \frac{1}{20} + \frac{1}{20} + \frac{1}{20} + \frac{1}{10}$

   Hundredths: _____

7. Write your own coin problem in the space below. Have your partner solve your problem.

   _____

   _____

   _____

   _____

LESSON 10·2 Finding Lines of Reflection

Dart Game

Practice before you play the game on Activity Sheet 7. One partner chooses Dart A and the other partner Dart B. Try to hit the target with your own dart, using the transparent mirror. **Do not practice with your partner's dart.**

Now play the game with your partner.

Directions

Take turns. When it is your turn, use the other dart—the one you did not use for practice. Try to hit the target by placing the transparent mirror on the page, but **do not look through the mirror.** Then both you and your partner look through the mirror to see where the dart hit the target. Keep score.

Pocket-Billiards Game

Practice before you play the game on Activity Sheet 8. Choose a ball (1, 2, 3, or 4) and a pocket (A, B, C, D, E, or F). Try to get the ball into the pocket, using the transparent mirror.

Now play the game with a partner.

Directions

Take turns. When it is your turn, say which ball and which pocket you have picked: for example, "Ball 2 to go into Pocket D." Try to get the ball into the pocket by placing the transparent mirror on the billiard table, **but do not look through the mirror.** Then both you and your partner look through the mirror to check whether the ball has gone into the pocket.

1. How could measuring distances with a ruler help you place the mirror so that the ball goes into the pocket? For example, exactly where can you put the mirror so that Ball 2 will go into Pocket D?

_____

_____

_____

LESSON 10·2 Math Boxes

1. Complete the table with equivalent names.

Fraction	Decimal	Percent
		29%
	0.30	
$\frac{8}{10}$		
		90%

SRB
61 62

2. Insert the decimal point in each quotient.

a. **2 8 2** $= 84.6 \div 3$

b. $91.6 \div 4 =$ **2 2 9**

c. **2 1 4** $= 128.4 \div 6$

d. $265.6 \div 8 =$ **3 3 2**

SRB
22 23

3. The fourth-grade students at Lighthouse School voted on their favorite season.

winter: 30 students

spring: 20 students

summer: 35 students

fall: 15 students

Use this data to create a bar graph.

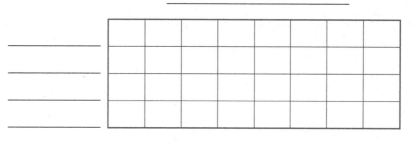

SRB
76

4. Write the ordered pair for each point plotted on the coordinate grid.

A (___ , ___)

B (___ , ___)

C (___ , ___)

D (___ , ___)

E (___ , ___)

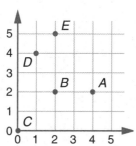

SRB
144

5. Write five names for −12.

a. _____

b. _____

c. _____

d. _____

e. _____

SRB
60

LESSON 10·3

Growing Patterns

1. a. Draw the next shape in the pattern.

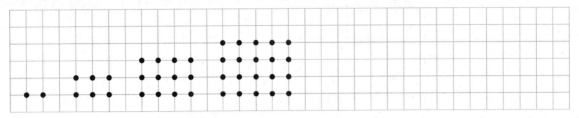

b. Describe the rule you used to extend the pattern.

_____

2. a. Extend the number pattern: 2, 6, 12, 20, _____, _____, _____

b. Describe the rule you used to extend the pattern.

_____

_____

3. a. Draw the next shape in the pattern.

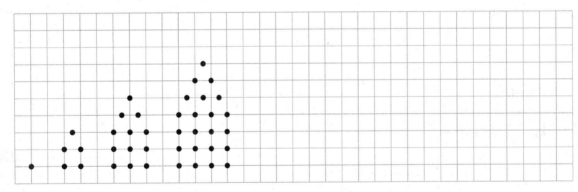

b. Describe the rule you used to extend the pattern.

_____

_____

4. a. Extend the number pattern: 1, 5, 12, 22, _____, _____, _____

b. Describe the rule you used to extend the pattern.

_____

_____

LESSON 10·3 Growing Patterns *continued*

5. a. Draw the next shape in the pattern.

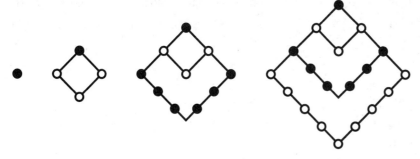

b. Describe the rule used to extend the pattern.

_____

_____

6. a. Extend the number pattern: 1, 4, 10, 19, _____, _____, _____

b. Describe the rule used to extend the pattern.

_____

_____

Try This

The numbers in Problem 6 are called *tetragonal numbers.* The shapes in Problem 5 show tetragonal numbers as arrays of dots.

7. a. Make a list of the first 20 tetragonal numbers. Circle the ones digit in each number.

①, ④, 1⓪, 1⑨, _____, _____, _____, _____, _____, _____, _____, _____, _____,

_____, _____, _____, _____, _____, _____, _____

b. Make a list of the ones digits in the first 20 tetragonal numbers.

1, 4, 0, 9, ___, ___, ___, ___, ___, ___, ___, ___, ___, ___, ___, ___, ___, ___, ___, ___

c. Describe any patterns you see in the list of numbers in Part b.

_____

_____

d. Do you think this pattern will continue? _____ Test your answer.

LESSON 10·3   **Math Boxes**

---

1. a. If you spin this spinner 400 times, about how many times would you expect it to land on

red? _____ times

blue? _____ times

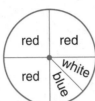

b. Explain how you figured out how many times you would expect the spinner to land on blue.

_____

_____

_____

_____

SRB
82–86

---

2. Complete.

Rule: _____

in	out
$\frac{3}{8}$	$\frac{7}{8}$
$\frac{1}{6}$	
	$\frac{9}{10}$
$\frac{7}{20}$	
$\frac{1}{4}$	$\frac{3}{4}$

SRB
55 57

3. Solve each open sentence.

a. $32.5 + y = 37.6$ $y =$ _____

b. $123.5 - k = 102.2$ $k =$ _____

c. $b + 67.91 = 78.32$ $b =$ _____

d. $405.08 - w = 231.57$ $w =$ _____

SRB
34–37

---

4. Angle *ART* is an _____ (acute or obtuse) angle.

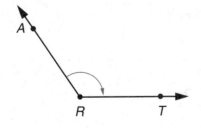

Measure of $\angle ART =$ _____ °

SRB
93
142 143

5. Alberto and Lara estimated the weight of their adult cat. What is the most reasonable estimate? Fill in the circle next to the best answer.

○ **A.** 2 pounds

○ **B.** 10 pounds

○ **C.** 50 pounds

○ **D.** 100 pounds

SRB
140

LESSON 10·4 **Line Symmetry**

You will need *Math Masters,* pages 311–314.

1. The drawings on *Math Masters,* page 311 are only half-pictures. Figure out what each whole picture would show. Then use a transparent mirror to complete each picture. Use the recessed side of the mirror to draw the line of reflection.

2. The pictures on *Math Masters,* page 312 are symmetric.

   a. Use the transparent mirror to draw the line of symmetry for the bat and the turtle.

   b. Cut out the other three pictures and find their lines of symmetry by folding.

   c. Which picture has two lines of symmetry? _____

3. Cut out each polygon on *Math Masters,* pages 313 and 314. Find all the lines of symmetry for each polygon. Record the results below.

Polygon	Number of Lines of Symmetry
A	
B	
C	
D	
E	

Polygon	Number of Lines of Symmetry
F	
G	
H	
I	
J	

4. Study the results in the tables above.

   a. How many lines of symmetry are in a regular pentagon (Polygon I)?   _____ lines

   b. How many lines of symmetry are in a regular hexagon (Polygon J)?   _____ lines

   c. How many lines of symmetry are in a regular octagon? (An octagon has 8 sides.)   _____ lines

LESSON 10·4 **Math Boxes**

1. Use your Geometry Template to draw the image of the figure that is shown above the line of reflection.

preimage

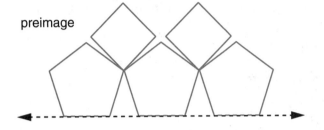

image

106

2. Use the following list of numbers to answer the questions.

7, 8, 24, 8, 9, 17, 17, 8, 12, 13, 19

a. What is the median? _____

b. Mode? _____

c. Maximum? _____

d. Minimum? _____

e. Range? _____

73

3. Order the fractions from least to greatest.

a. $\frac{8}{9}, \frac{4}{9}, \frac{1}{9}, \frac{9}{9}, \frac{3}{9}$ _____

b. $\frac{4}{4}, \frac{4}{9}, \frac{4}{12}, \frac{4}{3}, \frac{4}{2}$ _____

c. $\frac{3}{7}, \frac{1}{2}, \frac{7}{8}, \frac{1}{5}, \frac{4}{4}$ _____

53 54

4. Insert the decimal point in each product.

a. $4 * 6.7 =$ 2 6 8

b. 1 4 5 6 $= 28 * 5.2$

c. $7.3 * 46 =$ 3 3 5 8

d. 2 2 2 5 $= 0.25 * 89$

18 19

5. Jaleel missed 20% of the 30 problems on his science test. How many problems did he miss? Fill in the circle next to the best answer.

○ **A.** 5

○ **B.** 20

○ **C.** 3

○ **D.** 6

38
39 62

6. What are the perimeter and area of the rectangle? Include the correct units.

22 m

15 m

Perimeter = _____

Area = _____

131 134

LESSON 10·4 — Multiplying Fractions by Whole Numbers

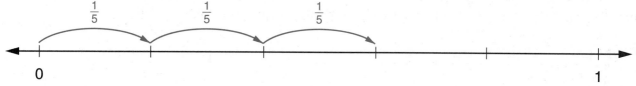

Write an equation to describe each number line.

1.

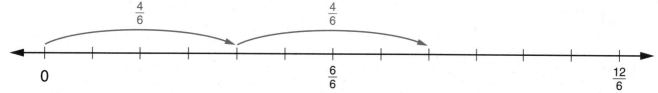

Equation: _____ * _____ = _____

2.

Equation: _____ * _____ = _____

Use the number lines to help you solve the problems.

3. _____ = $4 * \frac{2}{5}$

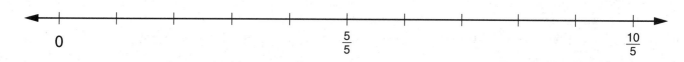

4. $3 * \frac{2}{8} =$ _____

Solve. You may draw a visual fraction model such as a number line if you wish.

5. $6 * \frac{1}{7} =$ _____   **6.** _____ = $8 * \frac{2}{10}$

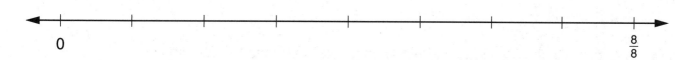

7. _____ = $5 * \frac{3}{12}$ **8.** $4 * \frac{2}{9} =$ _____

280A

LESSON 10·4 Solving Number Stories

A manufacturer of dry puppy food has the following feeding guidelines. All measurements are given in cups per day. Use the information in the table to answer the questions below. Write an equation to show what you did.

Weight (lb)	6–11 Weeks	3–4 Months	5–7 Months	8–12 Months
1	$\frac{1}{2}$	$\frac{1}{3}$	$\frac{1}{4}$	$\frac{1}{4}$
3	1	$\frac{3}{4}$	$\frac{1}{2}$	$\frac{1}{3}$
5	$\frac{4}{3}$	$\frac{5}{4}$	$\frac{3}{4}$	$\frac{1}{2}$
10	2	2	$\frac{5}{4}$	$\frac{2}{3}$
15	$\frac{11}{4}$	$\frac{11}{4}$	$\frac{5}{3}$	1
20	$\frac{10}{3}$	$\frac{13}{4}$	2	1

1. a. Buddy weighs 3 pounds and is 9 months old. According to the guidelines, he should eat about $\frac{1}{3}$ cup of food per day. How much food should Buddy eat in 5 days?

   _____ cups   Equation: _____

   b. Buddy should eat between _____ cups of food in 5 days. Circle the best answer.

   1 and 2   2 and 3   3 and 4

2. a. Cody is 8 weeks old and weighs about 5 pounds. How much food should he eat in 4 days?

   _____ cups   Equation: _____

   b. Cody should eat between _____ cups of food in 4 days. Circle the best answer.

   3 and 4   4 and 5   5 and 6

3. a. A puppy weighs 5 pounds and is 6 months old. How much food should the puppy eat in one week?

   _____ cups   Equation: _____

   b. The puppy should eat between _____ cups of food in one week. Circle the best answer.

   3 and 4   4 and 5   5 and 6

LESSON 10·5 **Frieze Patterns**   SRB 108

1. Extend the following frieze patterns. Use a straightedge and your transparent mirror to help you. Then write if you used a reflection, rotation, or translation of the original shape to continue the pattern.

   a. _____

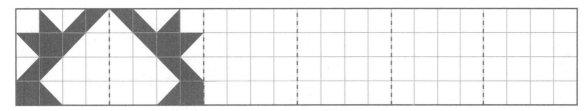

   b. _____

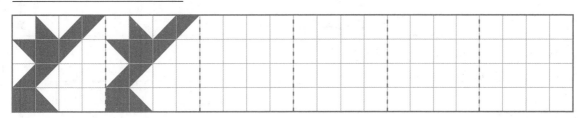

   c. _____

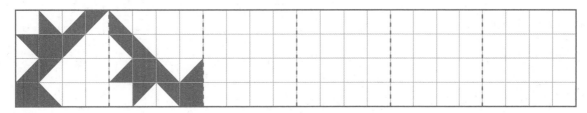

2. Create your own frieze pattern. Make a design in the first box. Then repeat the design, using reflections, slides, rotations, or a combination of moves. When you have finished, you may want to color or shade your frieze pattern.

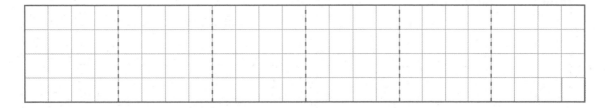

3. Explain how you created your pattern in Problem 2.

   _____

   _____

LESSON 10·5   # Math Boxes

---

1. Complete the table with equivalent names.

Fraction	Decimal	Percent
		31%
	0.10	
$\frac{4}{16}$		
		5%

61 62

2. Insert the decimal point in each quotient.

a. 74.8 ÷ 4 = **1 8 7**

b. **6 9** = 34.5 ÷ 5

c. 88.5 ÷ 3 = **2 9 5**

d. **2 4 2** = 193.6 ÷ 8

SRB
22 23

---

3. Julia babysat for the family next door. Below are the hours she worked.

Monday: 4 hours

Tuesday: $2\frac{1}{2}$ hours

Wednesday: 1 hour

Thursday: $1\frac{1}{2}$ hours

Friday: 3 hours

Use this data to create a bar graph.

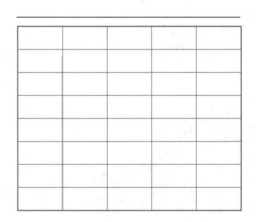

___ ___ ___ ___ ___

_____

76

---

4. Write the ordered pair for each point plotted on the coordinate grid.

A (___ , ___)

B (___ , ___)

C (___ , ___)

D (___ , ___)

E (___ , ___)

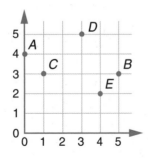

SRB
144

5. Write five names for −73.

a. _____

b. _____

c. _____

d. _____

e. _____

60

LESSON 10·6   **Review: Fractions, Decimals, and Percents**

1. Fill in the missing numbers in the table of equivalent fractions, decimals, and percents.

Fraction	Decimal	Percent
$\frac{4}{10}$		
	0.6	
		75%

2. Kendra set a goal of saving $50 in 8 weeks. During the first 2 weeks, she was able to save $10.

   a. What fraction of the $50 did she save in the first 2 weeks? _____

   b. What percent of the $50 did she save? _____

   c. At this rate, how long will it take her to reach her goal? _____ weeks

3. Shade 80% of the square.

   a. What fraction of the square did you shade? _____

   b. Write this fraction as a decimal. _____

   c. What percent of the square is *not* shaded? _____

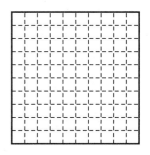

4. Tanara's new skirt was on sale at 15% off the original price. The original price of the skirt was $60.

   a. How much money did Tanara save with the discount? _____

   b. How much did she pay for the skirt? _____

5. Star Video and Vic's Video Mart sell videos at about the same regular prices. Both stores are having sales. Star Video is selling its videos at $\frac{1}{3}$ off the regular price. Vic's Video Mart is selling its videos at 25% off the regular price. Which store has the better sale? Explain your answer.

_____

_____

_____

LESSON
10·6
Math Boxes

1. a. Make a spinner. Color it so that if you spin it 36 times, you would expect it to land on blue 27 times and red 9 times.

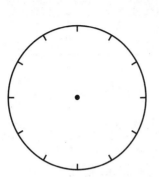

b. Explain how you designed your spinner.

_____

_____

_____

_____

SRB
82–86

2. Complete.

Rule: _____

in	out
$\frac{8}{16}$	$\frac{4}{16}$
$\frac{8}{8}$	
$\frac{3}{4}$	$\frac{2}{4}$
	$\frac{7}{12}$
$\frac{15}{20}$	

SRB
55 57

3. Solve each open sentence.

a. $67.3 + p = 75.22$ $p =$ _____

b. $6.86 - a = 2.94$ $a =$ _____

c. $x + 5.69 = 7.91$ $x =$ _____

d. $4.6 - n = 0.32$ $n =$ _____

SRB
34–37

4. Angle *RUG* is an _____ (acute or obtuse) angle.

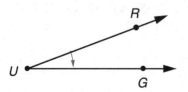

Measure of $\angle RUG =$ _____ °

SRB
93 142
143

5. Sebastian and Joshua estimated the weight of their mother. What is the most reasonable estimate? Fill in the circle next to the best answer.

○ **A.** 50 pounds

○ **B.** 150 pounds

○ **C.** 500 pounds

○ **D.** 1,000 pounds

SRB
140

Date _____   Time _____

---

1. What is the area of the rectangle?

9 cm

4.8 cm

Area = _____ cm²

134

2. Write five names for −214.

a. _____

b. _____

c. _____

d. _____

e. _____

60

---

3. For each animal, circle the most reasonable estimate of its weight.

a. raccoon   > 500 pounds   < 500 pounds   about 500 pounds

b. tiger   > 500 pounds   < 500 pounds   about 500 pounds

c. blue whale   > 500 pounds   < 500 pounds   about 500 pounds

d. giraffe   > 500 pounds   < 500 pounds   about 500 pounds

e. squirrel   > 500 pounds   < 500 pounds   about 500 pounds

---

4. Draw 3 different rectangles. Each should have an area of 12 square centimeters. Next to each rectangle, record its perimeter.

SRB
131 134

LESSON 11·1 Estimating Weights in Grams and Kilograms

A nickel weighs about 5 grams (5 g).
A liter of water weighs about 1 kilogram (1 kg).

In Problems 1–7, circle a possible weight for each object.

1. A dog might weigh about

    20 kg   200 kg   2,000 kg

2. A can of soup might weigh about

    4 g   40 g   400 g

3. A newborn baby might weigh about

    3 kg   30 kg   300 kg

4. An adult ostrich might weigh about

    1.5 kg   15 kg   150 kg

5. A basketball might weigh about

    0.6 kg   6 kg   60 kg

6. The weight limit in an elevator might be about

    100 kg   1,000 kg   10,000 kg

7. A pencil might weigh about

    4.5 g   45 g   450 g

8. Choose one of the problems above. Explain why you chose your answer.

_____

_____

LESSON 11·1 Metric and Customary Weight

The number line below has ounces on the top and grams on the bottom.
It shows, for example, that 7 ounces are about equal to 200 grams.

ounces

```
0     2     4     6     8    10    12    14    16    18    20
```

```
0        100      200        300       400       500
```

grams

Use the number line to give the approximate weight of each object.

1.

15 ounces

About _____ grams

2.

16 ounces

About _____ grams

3.

100 grams

About _____ ounces

4.

500 grams

About _____ ounces

Use the number line to help you determine which object weighs more. Circle the heavier object.

5.

225 grams

2.3 ounces

6.

5 ounces

454 grams

287

LESSON 11·1

Math Boxes

1. a. Explain how you know that the pattern below is an example of a translation.

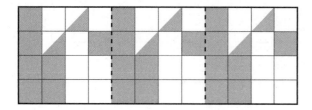

_____

_____

_____

b. Draw the figure after it is translated to the right.

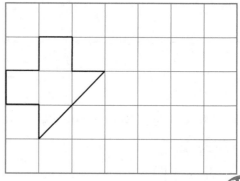

SRB
107

---

2. Find the solution of each open sentence.

a. $\frac{6}{7} - y = \frac{4}{7}$ $y =$ _____

b. $\frac{3}{10} + a = \frac{9}{10}$ $a =$ _____

c. $\frac{3}{5} - r = \frac{1}{10}$ $r =$ _____

d. $\frac{3}{4} + m = \frac{7}{8}$ $m =$ _____

SRB
55 148

---

3. Circle the numbers that are multiples of 6. Put an X through the numbers that are multiples of 5.

38

84

150

198

540

3,500

SRB
9

---

4. Insert parentheses to make each number sentence true.

a. $14 * 18 - 15 = 42$

b. $13 - 6 * 5 = 56 - 21$

c. $48 / 6 + 2 = 10 - 4$

d. $150 / 10 + 5 < 4 * 4$

SRB
150

---

5. If you use an average of 7 sheets of paper per day, about how many sheets would you use in

a. 1 week? _____ sheets

b. 4 weeks? _____ sheets

c. 52 weeks? _____ sheets

d. 2 years? _____ sheets

SRB
47

LESSON 11·2 **Geometric Solids**

Geometric shapes like these 3-dimensional ones are also called **geometric solids**.

Rectangular   Cylinder   Triangular   Cone   Sphere   Square
Prism   Prism   Pyramid

Look around the classroom. Try to find examples of the geometric solids pictured above. Draw a picture of each. Then write its name (for example: book).

Example of rectangular prism:	Example of cylinder:	Example of triangular prism:
Name of object:	Name of object:	Name of object:
_____	_____	_____
Example of cone:	Example of sphere:	Example of square pyramid:
Name of object:	Name of object:	Name of object:
_____	_____	_____

LESSON 11·2 Modeling a Rectangular Prism

After you construct a rectangular prism with straws and twist-ties, answer the questions below.

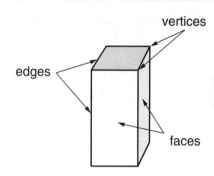

1. How many faces does your rectangular prism have? _____ face(s)

2. How many of these faces are formed by rectangles? _____ face(s)

3. How many of these faces are formed by squares? _____ face(s)

4. Pick one of the faces. How many other faces are parallel to it? _____ face(s)

5. How many edges does your rectangular prism have? _____ edge(s)

6. Pick an edge. How many other edges are parallel to it? _____ edge(s)

7. How many vertices does your rectangular prism have? _____ vertices

8. Write T (true) or F (false) for each of the following statements about the rectangular prism you made. Then write one true statement and one false statement of your own.

    a. _____ It has no curved surfaces.

    b. _____ All of the edges are parallel.

    c. _____ All of the faces are polygons.

    d. _____ All of the faces are congruent.

    e. True _____

    f. False _____

LESSON 11·2 **Making a 1-Ounce Weight**

SRB
140

1. Estimate how many of each coin you think it will take to make a 1-ounce weight. Then use a balance or scale to determine exactly how many of each coin are needed.

Coin	Estimated Number of Coins	Actual Number of Coins
penny		
nickel		
dime		
quarter		

2. Describe how you estimated how many of each coin it might take to make a 1-ounce weight.

_____

_____

_____

_____

_____

_____

_____

Try This

3. About what fraction of an ounce does each coin weigh?

1 penny = _____ oz   1 nickel = _____ oz   1 dime = _____ oz   1 quarter = _____ oz

Explain how you found your answers.

_____

_____

_____

LESSON 11·2   Math Boxes

---

1. The object below has the shape of a geometric solid. What is the name of the solid? Circle the best answer.

  A. rectangular prism

  B. cone

  C. cylinder

  D. square pyramid

101 102

2. Draw the figure after it is rotated clockwise $\frac{1}{4}$-turn.

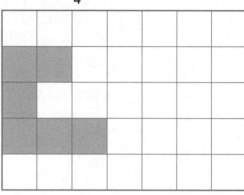

106 107

---

3. Write a number model to estimate the answer. Then correctly place the decimal point.

  a. 0.97 * 4 = 3 8 8

   Number model: _____

  b. 1 8 7 = 74.8 ÷ 4

   Number model: _____

4. Insert <, >, or = to make a true number sentence.

  a. −12 _____ −19

  b. −44 _____ 26

  c. −64 _____ −0.43

  d. $-\frac{1}{2}$ _____ $-\frac{4}{8}$

  e. −0.28 _____ −0.37

6   60

---

5. Round each number to the nearest tenth.

  a. 2.34 _____

  b. 0.68 _____

  c. 14.35 _____

  d. 1.62 _____

  e. 5.99 _____

182 183

6. A cinnamon raisin bagel has about 230 calories. How many calories are in one dozen bagels?

  About _____ calories

47

LESSON 11·3 Construction of Polyhedrons

Polyhedrons are geometric solids with flat surfaces formed by polygons.

For each problem below—

◆ Decide what the polyhedron should look like.

◆ Use straws and twist-ties to model the polyhedron.

◆ Answer the questions about the polyhedron.

Look at page 102 of the *Student Reference Book* if you need help with the name.

1. I am a polyhedron.
   I have 5 faces.
   Four of my faces are formed by triangles.
   One of my faces is a square.

   a. After you make me, draw a picture of me in the space to the right.

   b. What am I? _____

   c. How many corners (vertices) do I have? _____

   d. What shape is my base? _____

2. I am a polyhedron.
   I have 4 faces.
   All of my faces are formed by equilateral triangles.
   All of my faces are the same size.

   a. After you make me, draw a picture of me in the space to the right.

   b. What am I? _____

   c. How many corners (vertices) do I have? _____

   d. What shape is my base? _____

293

LESSON 11·3 Drawing a Cube

Knowing how to draw is a useful skill in mathematics. Here are a few ways to draw a cube. Try each way. Tape your best work at the bottom of page 295.

A Basic Cube

Draw a square.

Draw another square that overlaps your first square.
The second square should be the same size as the first.

Connect the corners of your 2 squares as shown.
This picture does not look much like a real cube. One problem is that the picture shows all 12 edges, even though not all the edges of a real cube can be seen at one time. Another problem is that it is hard to tell which face of the cube is in front.

A Better Cube

Begin with a square.

Next, draw 3 parallel line segments going right and up from 3 corners of your square. The segments should all be the same length.

Finally, connect the ends of the 3 line segments.

This cube is better than before, but it shows only the edges and corners, not the faces. If you want, try shading your cube to make it look more realistic.

LESSON 11·3   **Drawing a Cube** *continued*

A Cube with Hidden Edges

Sometimes people draw cubes and other shapes with dashed line segments.
The dashed line segments show edges that are hidden. Here is one way to draw
a cube with hidden edges. Use a pencil.

Draw a square.

Draw a faint square that overlaps your first square.
The second square should be the same size as the first.

Connect the corners of your 2 squares with faint line segments.

Trace over 5 of your faint line segments with solid lines
and 3 with dashed lines. The dashed line segments show the
3 edges that are hidden.

Tape your best work here.

Designing a Bookcase

Stephen wants to build a bookcase for his books. To help him design the bookcase, he measured the height of each of his books. He rounded each measurement to the nearest $\frac{1}{8}$ of an inch. His measurements are given below.

Book Heights (to the nearest $\frac{1}{8}$ inch)

$6\frac{1}{2}$, $9\frac{1}{4}$, $7\frac{1}{8}$, $7\frac{1}{2}$, 8, $6\frac{7}{8}$, $9\frac{1}{4}$, $9\frac{1}{4}$, $9\frac{1}{4}$, $9\frac{1}{4}$, $9\frac{1}{4}$, $8\frac{1}{4}$, 8, $8\frac{1}{4}$, $8\frac{3}{8}$,

$6\frac{1}{2}$, $7\frac{1}{8}$, 9, $6\frac{7}{8}$, $9\frac{3}{8}$, $6\frac{7}{8}$, $7\frac{1}{2}$, 8, $8\frac{1}{4}$, $9\frac{1}{4}$, $6\frac{7}{8}$, $6\frac{7}{8}$, $8\frac{1}{4}$, $8\frac{1}{4}$, $8\frac{1}{4}$

Plot the data set on the line plot below.

Book Heights

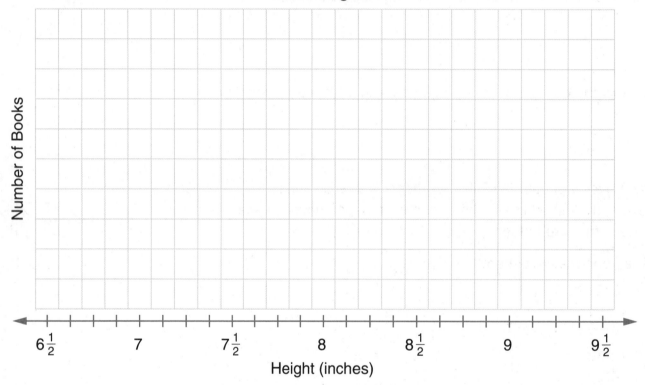

Use the completed plot to answer the questions below and on journal page 295B. Write a number model to show how you solved each problem.

1. a. What is the maximum book height? _____ inches

   b. What is the minimum book height? _____ inches

   c. What is the range of the data set? _____ inches

   Number model: _____

LESSON 11·3   **Designing a Bookcase** *continued*

SRB
55

2. **a.** What is the median of the data set? _____ inches

   b. How much longer is the maximum height than the median height?

   _____ inches   Number model: _____

3. Suppose that Stephen wants to make the space between the shelves on his bookcase $\frac{7}{8}$ of an inch taller than his tallest book.

   a. How far apart should he make the shelves?

   _____ inches apart   Number model: _____

   b. If the thickness of the wood he uses for the shelves is $\frac{5}{8}$ inch, what will be the total height of each shelf? (*Hint:* The total height is the thickness of the wood plus the distance between shelves.)

   _____ inches   Number model: _____

4. Stephen decides to make the bookshelf two shelves high. He will put all the books that are 8 inches tall or shorter on the top shelf and all the books that are more than 8 inches tall on the bottom shelf.

   a. What will be the difference in height between the tallest books on the top shelf and the shortest books on the top shelf?

   _____ inches   Number model: _____

   b. What will be the difference in height between the tallest book on the bottom shelf and the shortest books on the bottom shelf?

   _____ inches   Number model: _____

5. Make up and solve your own problem about the book height data.

   _____

   _____

   _____

   Number model: _____

LESSON
11·3   **Math Boxes**

---

1. Draw the figure after it is translated to the right.

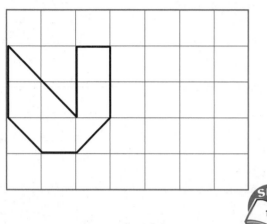

SRB
107

2. Find the solution of each open sentence.

a. $\frac{7}{8} - s = \frac{1}{8}$ $s =$ _____

b. $t + \frac{1}{4} = \frac{1}{2}$ $t =$ _____

c. $\frac{3}{10} - m = \frac{1}{5}$ $m =$ _____

d. $\frac{2}{8} + x = \frac{3}{4}$ $x =$ _____

SRB
55   148

---

3. Name the first ten multiples of each number.

a. 6 _____, _____, _____, _____, _____, _____, _____, _____, _____

b. 86 _____, _____, _____, _____, _____, _____, _____, _____, _____

SRB
9

---

4. Insert parentheses to make each number sentence true.

a. $98.3 + 1.7 * 2.5 = 250$

b. $21.7 / 3 + 4 = 3.1$

c. $56.3 + 3.7 * 3 > 5 * 30$

d. $13.8 - 8.3 = 26.15 - 23.4 * 2$

SRB
150

5. Gum costs $0.80 per pack. What is the cost of

a. 4 packs of gum? _____

b. 10 packs of gum? _____

c. 16 packs of gum? _____

d. 33 packs of gum? _____

LESSON 11·4   **Math Boxes**

---

1. The object below has the shape of a geometric solid. Name the solid.

_____

SRB
101

2. Draw the figure after it is rotated counterclockwise $\frac{1}{4}$-turn.

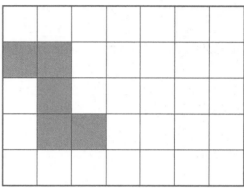

SRB
106 107

---

3. Write a number model to estimate the answer. Then correctly place the decimal point.

a. $7.56 * 4 = 3\ 0\ 2\ 4$

Number model: _____

b. $563.2 \div 4 = 1\ 4\ 0\ 8$

Number model: _____

4. Insert $<$, $>$, or $=$ to make a true number sentence.

a. -14 _____ -6

b. -123 _____ -241

c. -8.9 _____ -5.7

d. $-\frac{1}{4}$ _____ $-\frac{2}{5}$

e. $-\frac{3}{9}$ _____ $-\frac{1}{3}$

SRB
6 60

---

5. Round each number to the nearest tenth.

a. 3.46 _____

b. 0.71 _____

c. 4.35 _____

d. 9.60 _____

e. 22.89 _____

SRB
182 183

6. Jake can ride his bike 5 miles in 40 minutes. At this rate, how long does it take him to ride 1 mile? Circle the best answer.

A. 200 minutes

B. 40 minutes

C. 20 minutes

D. 8 minutes

LESSON 11·5

Area of a Rectangle

1. Write a formula for the area of a rectangle. In your formula, use A for area. Use l and w for length and width, or b and h for base and height.

_____

2. Draw a rectangle with sides measuring 3 centimeters and 9 centimeters. Find the area.

Number model: _____   Area = _____ square centimeters

3. Find the height of the rectangle.

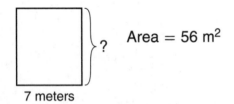

Area = 56 m²

7 meters   ?

Number model: _____

height = _____ m

4. Find the length of the base of the rectangle.

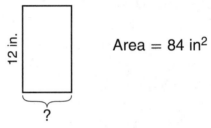

12 in.   Area = 84 in²

?

Number model: _____

length of base = _____ in.

Try This

5. Find the area of the rectangle.

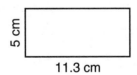

5 cm

11.3 cm

Number model: _____

Area = _____ cm²

6. Find the height of the rectangle.

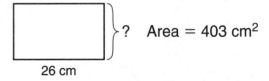

? Area = 403 cm²

26 cm

Number model: _____

height = _____ cm

LESSON 11·5   Math Boxes

---

1. What is the total number of cubes needed to completely fill the box?

_____ cubes

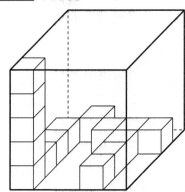

SRB
138

---

2. Calculate the volume.

35 in

35 in.

25 in.

Number model: _____

Volume = _____ in³

SRB
138

---

3. When you roll a 6-sided die, about what fraction of the time would you expect

a. a multiple of 2
to come up?   _____

b. a factor of 20
to come up?   _____

SRB
81

---

4. Complete.

a. 13 ft = _____ yd _____ ft

b. 18 ft 6 in. = _____ yd _____ in.

c. 972 in. = _____ yd

d. 15,840 ft = _____ mi

e. 24,640 yd = _____ mi

SRB
129

---

5. Add.

a. $-54 + 28 =$ _____

b. $-62 + (-15) =$ _____

c. _____ $= 51 + (-139)$

d. _____ $= -\$23.56 + \$87.45$

e. $\$71.08 + (-\$85.79) =$ _____

---

6. If 4 shirts cost $76, what is the cost of

a. 2 shirts? _____

b. 6 shirts? _____

c. 1 dozen shirts? _____

d. 75 shirts? _____

SRB
47

LESSON 11·5 Cube-Stacking Problems

Each picture at the bottom of this page and on the next page shows a box that is partially filled with cubes. The cubes in each box are the same size. Each box has at least one stack of cubes that goes to the top.

Your task is to find the total number of cubes needed to completely fill each box.

Record your answers in the table below.

Table of Volumes						
Placement of Cubes	**Box 1**	**Box 2**	**Box 3**	**Box 4**	**Box 5**	**Box 6**
Number of cubes needed to cover the bottom						
Number of cubes in the tallest stack (Be sure to count the bottom cube.)						
Total number of cubes needed to fill the box						

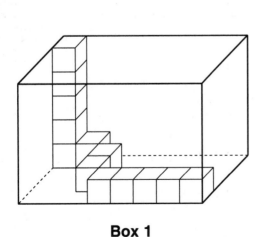

Box 1

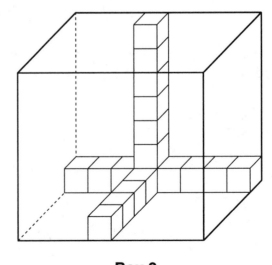

Box 2

LESSON 11·5 Cube-Stacking Problems *continued*

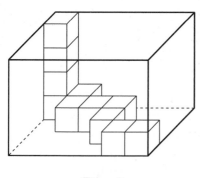

Box 3

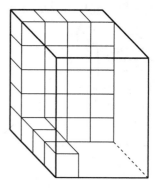

Box 4

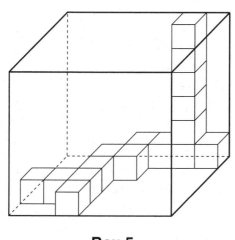

Box 5

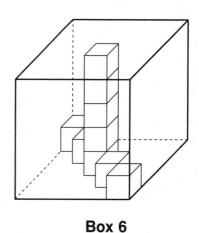

Box 6

Formula for the volume of a rectangular prism:

_____

B is the **area** of a base.

h is the height from that base.

Volume units are cubic units.

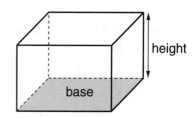

Date _____ Time _____

Find the volume of each stack of centimeter cubes.

1.

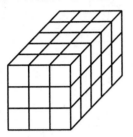

Volume = _____ cm³

2.

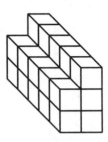

Volume = _____ cm³

3.

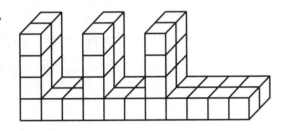

Volume = _____ cm³

4.

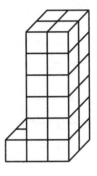

Volume = _____ cm³

5. Choose one of the problems from above. Describe the strategy that you used to find the volume of the stack of centimeter cubes.

_____

_____

_____

Try This

6.

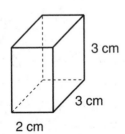

3 cm

3 cm

2 cm

Number model: _____

Volume = _____ cm³

7.

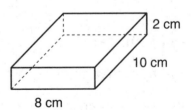

2 cm

10 cm

8 cm

Number model: _____

Volume = _____ cm³

LESSON 11·6 Gram and Ounce Museum

1. a. What was the heaviest item in the class Gram and Ounce Museum? _____

   b. How much did it weigh? _____ grams   _____ ounces

2. a. What was the lightest item in the class Gram and Ounce Museum? _____

   b. How much did it weigh? _____ grams   _____ ounces

Complete.

3. 6 g = _____ mg

4. _____ g = 7,000 mg

5. 3 kg = _____ g

6. _____ kg = 8,000 g

7. 2.9 g = _____ mg

8. _____ kg = 4,500 g

9. 6 lb = _____ oz

10. _____ lb = 144 oz

11. 3.5 lb = _____ oz

12. 8 T = _____ lb

Use the Rules of Thumb below to solve Problems 13–15. Write number models to show how you estimated.

> **Rules of Thumb**
>
> 1 kilogram equals about 2.2 pounds
> 1 ounce equals about 30 grams

13. A video camera weighs about 120 grams. About how many ounces is that?

   Number model: _____   _____ oz

14. A baby weighs about 3.5 kilograms at birth. About how many pounds is that?

   Number model: _____   _____ lb

15. An African elephant weighs 11,023 pounds. About how many kilograms is that?

   Number model: _____   _____ kg

Date _____   Time _____

LESSON 11·6   **Math Boxes**

---

1. The object below has the shape of a geometric solid. Name the solid.

_____

SRB 101

2. Which figure below shows the original figure rotated clockwise $\frac{1}{2}$-turn?

_____

Original   A   B   C

SRB 106 107

---

3. Write a number model to estimate the answer. Then correctly place the decimal point.

a. 6 * 32.9 = 1 9 7 4

Number model: _____

b. 3 2 9 = 98.7 ÷ 3

Number model: _____

4. Insert <, >, or = to make a true number sentence.

a. −34 _____ −9

b. −89 _____ −99

c. −2.99 _____ −2.9

d. $-\frac{1}{4}$ _____ $-\frac{1}{3}$

e. $-\frac{18}{9}$ _____ $-2\frac{1}{4}$

SRB 6 60

---

5. Round 8.99 to the nearest tenth. Circle the best answer.

A. 8.0

B. 9.0

C. 9.1

D. 8.09

SRB 182 183

6. It takes 2 cups of flour to make about 20 medium-size peanut butter cookies. How many cups of flour will you need to make about

a. 40 cookies? _____ cups

b. 60 cookies? _____ cups

c. 50 cookies? _____ cups

d. 740 cookies? _____ cups

SRB 47

LESSON 11·7 Measuring Capacity

SRB
137

Math Message

1 pint = _____ cups

1 quart = _____ pints

1 half-gallon = _____ quarts

1 gallon = _____ quarts

Think: How can the picture above help you remember how many cups are in a pint, how many pints are in a quart, and how many quarts are in a gallon?

Units of Capacity

1. Circle the unit you would use to measure each amount.

   A large jug of milk   milliliters or liters

   Water in a thimble   milliliters or liters

   A glass of juice   milliliters or liters

   Water in a water cooler   milliliters or liters

   Water in a fish tank   milliliters or liters

   Liquid in a paper cup   milliliters or liters

   A tank of gas   milliliters or liters

   A spoonful of oil   milliliters or liters

   A large bottle of water   milliliters or liters

   A can of soup   milliliters or liters

2. Explain how you decided which unit to use for a can of soup.

   _____

   _____

Date _____   Time _____

1. Shade in the appropriate amount to show the capacity of each of your containers.

a.

Container _____

b.

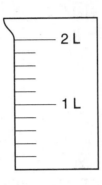

Container _____

c.

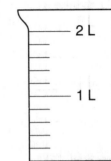

Container _____

d.

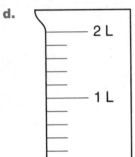

Container _____

e.

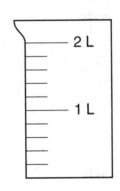

Container _____

f. Circle the container with the largest capacity. Was your prediction accurate?

_____

Units of Capacity	
U.S. Customary	**Metric**
1 gallon (gal) = 4 quarts (qt)	1,000 milliliter (mL) = 1 liter (L)
1 quart (qt) = 2 pints (pt)	1 milliliter (mL) = $\frac{1}{1,000}$ liter (L)
1 pint (pt) = 2 cups (c)	
1 pint (pt) = 16 fluid ounces (fl oz)	

2. Use the conversion table above to solve the problems.

   a. 6 qt = _____ pt

   b. _____ mL = 8 L

   c. _____ pt = 48 fl oz

   d. 6,450 mL = _____ L

   e. 10 qt = _____ gal

   f. _____ mL = 0.500 L

   g. 4 gal = _____ c

   h. 32 mL = _____ L

LESSON 11·7   # Solving Capacity Problems

Solve. You may draw pictures to help you.

1. Adaline filled her watering can with 1,250 mL of water.
    After watering her plants she had 485 mL left.
    How much water did she use?   _____ mL

2. Betty and Don spent the morning squeezing oranges
    for juice. Betty squeezed $1\frac{2}{4}$ L and Don squeezed $1\frac{3}{4}$ L.
    What is the total amount of juice?   _____ L

3. There are 450 mL of syrup in 1 can. What is the
    total amount of syrup in 6 cans?   _____ mL

4. Dimitra poured $\frac{2}{5}$ liter of water into a fish tank. William
    poured $\frac{4}{5}$ liter of water into the fish tank.

    a. How much more water did William pour?   _____ L

    b. How many milliliters is that?   _____ mL

5. Raina brought a 1,500 mL jug of water to the school
    picnic. Her water jug has enough water to fill 5 glasses.
    How much does each glass hold?   _____ mL

6. The teacher set out 24 bowls of glue for the students
    to use for an art project. Each bowl holds 75 mL of glue.
    How much glue did the teacher need to fill all the bowls? _____ mL

LESSON 11·7   # Largest Cities by Population

SRB
76 302

1. Use the data in the Largest Cities by Population table at the top of *Student Reference Book,* page 302 to complete the bar graph. Round each figure to the nearest million.

Largest Cities by Population

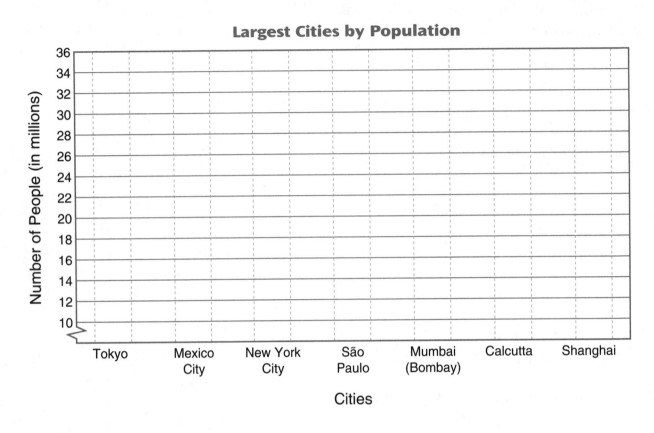

Cities

2. Make three statements comparing the cities in the bar graph.

Example: *About 21 million more people live in Tokyo than in Shanghai.*

_____

_____

_____

_____

LESSON 11·7 **Math Boxes**

1. What is the total number of cubes needed to completely fill the box?

_____ cubes

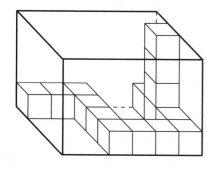

2. Calculate the volume.

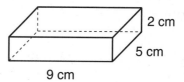

Number model: _____

Volume = _____ cm³

3. When you roll a 10-sided die, about what fraction of the time would you expect a multiple of 3 to come up?

_____

Use a probability term to describe the likelihood of this event.

_____

4. Complete.

a. 321 cm = _____ m

b. 56 cm = _____ mm

c. 14 ft 4 in. = _____ in.

d. 2 mi = _____ ft

e. 5.3 km = _____ m

f. _____ mi = 7,040 yd

5. Add.

a. $-46 + 20 =$ _____

b. $-23 + (-18) =$ _____

c. _____ $= 33 + (-17)$

d. _____ $= \$36.54 + (-\$57.81)$

e. $-\$131.09 + (-\$76.98) =$ _____

6. If you travel at an average speed of 50 miles per hour, how far will you travel in

a. 3 hours? _____ miles

b. $\frac{1}{2}$ hour? _____ miles

c. $2\frac{1}{2}$ hours? _____ miles

d. $5\frac{3}{5}$ hours? _____ miles

LESSON 11·8 | **Math Boxes**

---

1. If you use the telephone an average of 4 times per day, about how many times would you use it in

   a. 1 week? _____ times

   b. 4 weeks? _____ times

   c. 52 weeks? _____ times

---

2. A cup of orange juice has about 110 calories. About how many calories are in a quart of orange juice?

   _____ calories

---

3. Pears cost $0.55 each. What is the cost of

   a. 4 pears? _____

   b. 10 pears? _____

   c. 18 pears? _____

---

4. If you walk at an average speed of 3.5 miles per hour, how far will you travel in

   a. 2 hours? _____ miles

   b. 6 hours? _____ miles

   c. $\frac{1}{2}$ hour? _____ miles

---

5. Michelle can run 5 miles in 35 minutes. At this rate, how long does it take her to run 1 mile?

   _____ minutes

---

6. Round each number to the nearest tenth.

   a. 5.87 _____

   b. 0.32 _____

   c. 9.65 _____

   d. 3.40 _____

   e. 93.29 _____

LESSON 12·1 **Rates**

1. While at rest, a typical student in my class blinks _____ times in one minute.

2. While reading, a typical student in my class blinks _____ times in one minute.

3. In Problems 1 and 2, what is meant by the phrase *a typical student?*

   _____

   _____

4. Calculate the mean for each set of data.

   a. At rest: _____ blinks per minute

   b. While reading: _____ blinks per minute

5. List as many examples of rates as you can.

   _____

   _____

   _____

   _____

6. Find at least 2 examples of rates in your *Student Reference Book.*
   (*Hint:* Look at pages 271 and 299.)

   _____

   _____

   _____

   _____

LESSON 12·1   Counting with Fractions and Decimals

Fill in the missing fractions on the number lines below.

1.

2.

3.

Fill in the missing decimals on the number lines below.

4.

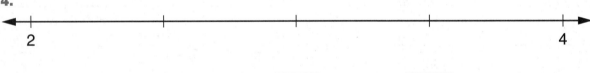

Try This

5.

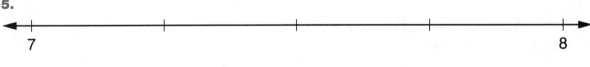

6.

LESSON 12·1 **Math Boxes**

1.

a. Pick a face of the cube. How many other faces are parallel to it?

_____ face(s)

b. Pick an edge of the cube. How many other edges are parallel to it?

_____ edge(s)

94 101

---

2. Calculate the volume.

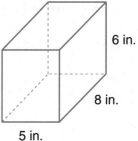
6 in.

8 in.

5 in.

Number model: _____

Volume = _____ in³

138

---

3. Write A, P, or V to tell whether you would need to find the area, perimeter, or volume in each situation.

a. Finding the distance around a circular track _____

b. Buying tile for a bathroom floor _____

c. Filling a pool with water _____

131 133
137

---

4. Insert <, >, or = to make a true number sentence.

a. −15 _____ 3

b. −43 _____ −21

c. 68 _____ −100

d. $-\frac{3}{4}$ _____ −0.78

e. $-13\frac{1}{2}$ _____ −13

60

---

5. For which number is 8 a factor? Fill in the circle next to the best answer.

Ⓐ 253

Ⓑ 94

Ⓒ 120

Ⓓ 884

7

---

6. Round each number to the nearest hundredth.

a. 12.368 _____

b. 234.989 _____

c. 1.225 _____

d. 12.304 _____

e. 0.550 _____

182 183

LESSON 12·2   # Rate Tables

For each problem, fill in the rate table. Then answer the question below the table.

1. Bill's new car can travel about 35 miles on 1 gallon of gasoline.

   Gasoline mileage: 35 miles per gallon

Miles	35							
Gallons	1	2	3	4	5	6	7	8

   At this rate, about how far can the car travel on 7 gallons of gas? _____ miles

2. Jennifer earns $8 for every 4 hours she helps out around the house.

   Earnings: $8 in 4 hours

Dollars				8				
Hours	1	2	3	4	5	6	7	8

   At this rate, how much money does Jennifer earn per hour? $_____

3. A gray whale's heart beats 24 times in 3 minutes.

   Gray whale's heart rate: 24 beats in 3 minutes

Heartbeats			24					
Minutes	1	2	3	4	5	6	7	8

   At this rate, how many times does a gray whale's heart beat in 2 minutes? _____ times

4. Ms. Romero paid $1.80 for 3 pounds of grapes.

   Cost of grapes: 3 pounds for $1.80

Pounds	1	2	3	4	5	6	7	8
Dollars			1.80					

   At this rate, how much do 5 pounds of grapes cost? $_____

LESSON 12·2 **Rate Tables** *continued*

5. Malia bought 6 yards of fabric for $15.00.

Cost of fabric: 6 yards for $15.00

Dollars						15.00		
Yards	1	2	3	4	5	6	7	8

At this rate, how much will $7\frac{1}{2}$ yards of fabric cost? $ _____

6. Alden bought $\frac{3}{4}$ of a pound of cheese for $6.

Cost of cheese: $\frac{3}{4}$ of a pound for $6

Pounds	$\frac{1}{4}$	$\frac{1}{2}$	$\frac{3}{4}$	1	$1\frac{1}{4}$	$1\frac{1}{2}$	$1\frac{3}{4}$	2
Dollars			6					

At this rate, how much will $1\frac{1}{4}$ pounds cost? $ _____

Try This

7. The Jefferson family plans to sit down to Thanksgiving dinner at 6:00 P.M. They have
   an 18-pound turkey. The turkey needs to cook about 20 minutes per pound.

   a. At what time should the turkey go in the oven? _____

   b. Explain what you did to solve the problem.

   _____

   _____

   _____

   _____

LESSON 12·2 **Math Boxes**

1. **a.** Complete the table.

Number of Pizzas	1	2		4	5
Number of Servings	3		9	12	

b. How many pizzas are needed for 279 servings?

_____ pizzas

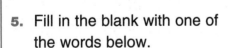

2. Complete.

a. 3 gal = _____ qt

b. 5 L = _____ mL

c. _____ L = 750 mL

d. _____ pt _____ c = 23 c

e. 64 gal = _____ pt

137

3. Find the solution of each open sentence.

a. $m + 40 = -60$ $m =$ _____

b. $55 + q = 40$ $q =$ _____

c. $(-23) + s = 0$ $s =$ _____

d. $p + (-36) = -80$ $p =$ _____

148

4. Complete the name-collection box.

8.01

149

5. Fill in the blank with one of the words below.

impossible

unlikely

likely

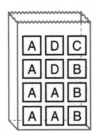

Without looking, it is _____ that a B block will be pulled from the bag.

80

6. Calculate.

a. 10% of 460 = _____

b. 5% of 120 = _____

c. 40% of _____ = 4

d. _____% of 20 = 6

e. _____% of 92 = 46

f. _____% of 150 = 57

38 39

LESSON 12·3

Math Boxes

1.

a. Pick a face of the cube. How many other faces are perpendicular to it?

_____ face(s)

b. Pick an edge of the cube. How many other edges are perpendicular

to it? _____ edge(s)

94 101

---

2. Calculate the volume.

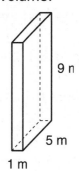

9 n

5 m

1 m

Number model: _____

Volume = _____ m³

138

---

3. Write A, P, or V to tell whether you would need to find the area, perimeter, or volume in each situation.

a. Buying paint for a bedroom ceiling _____

b. Buying a wedding ring _____

c. Buying dirt for a potted plant _____

131 133
137

---

4. Insert <, >, or = to make a true number sentence.

a. $8 _____ −$3

b. −$7 _____ −$2

c. $18 _____ −$11

d. $61.50 _____ −$67.85

e. −$203.90 _____ $320.10

60

---

5. Name all the factors of each number.

a. 55 _____

b. 32 _____

c. 96 _____

7

---

6. Round each number to the nearest hundredth.

a. 0.123 _____

b. 4.568 _____

c. 6.155 _____

d. 9.780 _____

e. 0.006 _____

182 183

Do These Numbers Make Sense?

Math Message

It is estimated that the average lifetime of a person living in the United States is about 75 years.

About how many days are there in an average lifetime? _____ days

About how many hours is that? _____ hours

Use the data from the above Math Message to help you answer the following questions:

1. It is estimated that a person sleeps about 214,000 hours in an average lifetime.

    a. At that rate, about how many hours *per day* does a person sleep? _____ hours per day

    b. Show or explain how you got your answer.

    c. Does this number make sense to you? Explain.

    _____

    _____

2. It is estimated that in an average lifetime a person watches about 105,000 hours of TV.

    a. At that rate, about how many hours *per day* does a person watch TV? _____ hours per day

    b. Show or explain how you got your answer.

    c. Does this number make sense to you? Explain.

    _____

    _____

LESSON 12·3   Do These Numbers Make Sense? *continued*

3. It is estimated that in an average lifetime a person laughs about 540,000 times.

    a. At that rate, about how many times *per day* does a person laugh? _____ times per day

    b. Show or explain how you got your answer.

    c. Does this number make sense to you? Explain.

    _____

    _____

    _____

4. It is estimated that in an average lifetime, a person takes about 95,000,000 breaths. Does this number make sense to you? Explain.

    _____

    _____

    _____

Try This

5. Write your own problem. Ask a partner to decide whether or not the numbers in your problem make sense.

    _____

    _____

    _____

LESSON 12·3 Line Graph

1. Use the data in the table below to create a line graph showing how the total amount of precipitation (rain and snow) changes from month to month in Ottawa, the capital of Canada.

   Use a straightedge to connect the points. Label each axis, and give the graph a title.

Month	J	F	M	A	M	J	J	A	S	O	N	D
Precipitation (in mm)	51	50	57	65	77	84	87	88	84	75	81	73

Source: www.theweathernetwork.com/weather/stats/pages/C01930.htm

Try This

2. In Ottawa, Canada, it rains or snows _____ mm during a typical month.

LESSON 12·4 Product Testing

Some publications ask their readers to test many different kinds of products. The results of the tests are then published to help readers make wise buying decisions. One example is the former *Consumer Reports for Kids Online*. It featured articles previously published in *Zillions,* a child's version of *Consumer Reports.* In one test, 99 readers field tested several backpack models. The readers considered fit, back friendliness, and comfort as they tried to decide which brand was the best buy. In another test, a team of young people compared 40 brands of jeans in their search for a brand that would not shrink in length.

When a reader wrote to complain about a board game she bought, the staff sent board games to young people in every part of the country. Testers were asked to play each game several times and then to report what they liked and disliked about the game.

1. If you were testing a board game, what are some of the features you would look for?

   _____

   _____

   _____

   _____

2. When readers of the magazine tested backpacks, they considered fit, back friendliness, and comfort in determining the best one. Which of these factors is the most important to you? Why?

   _____

   _____

   _____

   _____

3. What is a **consumer**? Be prepared to share your definition with the class.

   _____

   _____

   _____

   _____

LESSON 12·4 Converting Units of Measure

> 1 pound (lb) = 16 ounces (oz)
>
> 1 ton (T) = 2,000 pounds (lb)

Complete the conversions between ounces, pounds, and tons.
Write the information you needed to make the conversion.

1. 5 lb = _____ oz   *1 lb = 16 oz*

2. $\frac{1}{2}$ lb = _____ oz   _____

3. 129 lb = _____ oz   _____

4. 2 T = _____ lb   _____

5. $6\frac{1}{4}$ T = _____ lb   _____

6. 112 oz = _____ lb   _____

7. 20,000 lb = _____ T   _____

8. 4,240 oz = _____ lb   _____

9. 3,000 lb = _____ T   _____

10. $1\frac{1}{4}$ lb = _____ oz   _____

Complete the table.

11.

Ounces	Pounds	Tons
	10,000	
		9.5
16,000		

LESSON 12·4 **Converting Units of Measure** *continued*

SRB
315

12. Record measurement equivalents in the two-column tables below.

a.
Feet	Inches
1	12
2	
3	
4	
5	

b.
Liters	Milliliters
1	

c.
Minutes	Seconds
1	

Find the equivalent measures.

13. **a.** 6 km = _____ m

b. 2 L = _____ mL

c. 6 yd = _____ ft

d. 3.25 L = _____ mL

e. $5\frac{1}{2}$ kg = _____ g

f. $8\frac{1}{2}$ hr = _____ min

g. What do you notice when you convert from a larger unit to a smaller unit (such as from L to mL)?

_____

_____

14. **a.** 4,000 g = _____ kg

b. 200 cm = _____ m

c. 875 mL = _____ L

d. 660 sec = _____ min

e. 1,500 mL = _____ L

f. 156 in. = _____ ft

g. What do you notice when you convert from a smaller unit to a larger unit (such as from mL to L)?

_____

_____

319B

Date _____ Time _____

Solve the unit price problems below. Complete the tables if it is helpful to do so.

1. A 12-ounce can of fruit juice costs 60 cents. The unit price is _____ per ounce.

Dollars				0.60
Ounces	1	3	6	12

2. A 4-pound bunch of bananas costs $1.16. The unit price is _____ per pound.

Dollars				1.16
Pounds	1	2	3	4

3. A 5-pound bag of apples costs $1.90. The unit price is _____ per pound.

Dollars					1.90
Pounds	1	2	3	4	5

4. Three pounds of salmon cost $21.00.

   a. The unit price is _____ per pound.

   b. What is the cost of 7 pounds of salmon? _____

   c. What is the cost of $9\frac{1}{2}$ pounds of salmon? _____

Dollars			21.00			
Pounds	1	2	3	4	7	$9\frac{1}{2}$

Try This

5. *Energy* granola bars come in packages of 25 and cost $3.50 per package. *Super* granola bars come in packages of 30 and cost $3.60 per package. Which is the better buy? Explain.

_____

_____

_____

_____

LESSON 12·4

Math Boxes

---

1. a. Complete the table.

Number of Inches		72			540
Number of Yards	1	2	9	12	

b. How many inches are in 329 yards?

_____ inches

SRB
47

---

2. Complete.

a. 7 gal = _____ qt

b. 8 L = _____ mL

c. _____ L = 250 mL

d. _____ gal _____ qt = 25 qt

e. _____ qt _____ pt = 14 c

SRB
137

---

3. Find the solution of each open sentence.

a. $t + 30 = -120$ $t =$ _____

b. $75 + n = 20$ $n =$ _____

c. $16 + b = 0$ $b =$ _____

d. $c + (-61) = -97$ $c =$ _____

SRB
148

---

4. Which one of the names below is *not* a name for 3.16? Fill in the circle next to the best answer.

Ⓐ $4.8 - 1.64$

Ⓑ $15.8 / 5$

Ⓒ $2.47 * 6$

Ⓓ $2.5 + 0.66$

SRB
149

---

5. Add blocks to the bag so it is likely that Arjan will pick a C block without looking.

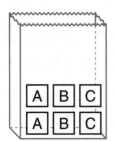

SRB
80

---

6. Calculate.

a. 10% of 520 = _____

b. 5% of 180 = _____

c. 40% of _____ = 4

d. _____% of 30 = 15

e. _____% of 35 = 14

f. _____% of 95 = 38

SRB
38 39

LESSON 12·5 Unit Pricing

SRB
47

Math Message

1. Use your calculator to divide. Write down what the calculator displays for each quotient. Your teacher will tell you how to fill in the answer spaces for "cents."

   a. $9.52 \div 7 = \$\rule{0.5cm}{0.4pt}.\rule{3cm}{0.4pt}$, or _____ cents

   b. $1.38 \div 6 = \$\rule{0.5cm}{0.4pt}.\rule{3cm}{0.4pt}$, or _____ cents

   c. $0.92 \div 8 = \$\rule{0.5cm}{0.4pt}.\rule{3cm}{0.4pt}$, or _____ cents

   d. $0.98 \div 6 = \$\rule{0.5cm}{0.4pt}.\rule{3cm}{0.4pt}$, or about _____ cents

   e. $1.61 \div 9 = \$\rule{0.5cm}{0.4pt}.\rule{3cm}{0.4pt}$, or about _____ cents

2. A package of 6 fruit bars costs $2.89. What is the price of 1 fruit bar? _____ cents

3. A 15-ounce bottle of shampoo costs $3.89. What is the price per ounce? _____ cents

4. Brand A: a box of 16 crayons for 80 cents
   Brand B: a box of 32 crayons of the same kind for $1.28

   Which box is the better buy? _____

   Why? _____

   _____

   _____

Try This

5. A store sells a 3-pound can of coffee for $7.98 and a 2-pound can of the same brand for $5.98. You can use a coupon worth 70 cents toward the purchase of the 2-pound can. If you use the coupon, which is the better buy, the 3-pound can or the 2-pound can? Explain your answer.

   _____

   _____

   _____

LESSON 12·5 **Investigating Liters and Milliliters**

Write the amount of liquid shown in liters.

1.

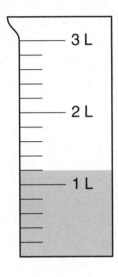

_____ liters

2.

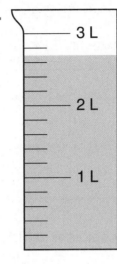

_____ liters

3.

_____ liters

Shade each pitcher to show the appropriate amount in milliliters.
Then record the amount in liters.

4. 2,400 mL = _____ L   5. 500 mL = _____ L   6. 1,950 mL = _____ L

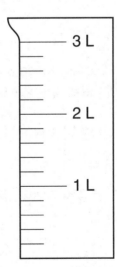

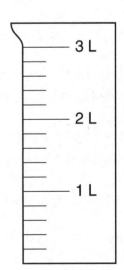

LESSON 12·5   **Investigating Liters and Milliliters** *continued*

Solve. You may draw pictures to help you.

Maria Elena's store sells soup in various sized containers.

Size of Container	Amount	Price
Extra small	250 mL	$1.50
Small	500 mL	$2.45
Medium	1 L	$4.80
Large	1.5 L	$6.25
Extra large	2 L	$8.50

1. How much more soup does the extra large container hold than the extra small?

    _____

2. Lucius bought 1 extra small, 1 medium, and 3 small
    containers of soup. How many liters of soup did he buy? _____ L

3. Kamu bought 3 large containers of soup. She needs 4,000 mL of soup.

    a. Does she have enough? _____

    b. If so, how much extra soup does she have? _____

4. Sally bought 1 extra small, 1 small, and 2 extra large
    containers of soup. How much money did she spend? _____

5. Rocco wants to spend $8 or less on soup but he wants to get
    the most for his money.

    a. What's the greatest amount of soup he could buy? _____

    b. How much would he spend? _____

6. Write and solve your own problem.

    _____

    _____

    _____

LESSON 12·5 Math Boxes

1. Multiply. Show your work.

$46 * 231 =$ _____

18 19

2. a. Complete the table.

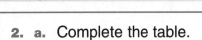

Number of Cups		32			272
Number of Gallons	1	2	9	12	

b. How many cups are in $5\frac{3}{4}$ gallons?

_____ cups

47

3. An average 10-year-old drinks about 20 *gallons* of soft drinks per year.

At that rate, about how many *cups* does a 10-year-old drink in a month?

_____

47

4. Subtract.

a. $+\$9 - (+\$4) =$ _____

b. $+\$8 - (-\$3) =$ _____

c. $-\$7 - (+\$15) =$ _____

d. $-\$6 - (+\$1) =$ _____

e. $-\$14 - (-\$9) =$ _____

60

5. Calculate the volume.

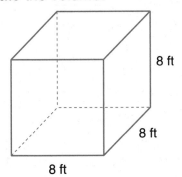

8 ft

8 ft

8 ft

Number model: _____

138

Volume = _____ ft^3

6. A 3-ounce bag of corn chips costs $0.65. A 14-ounce bag of corn chips costs $2.79.

a. What is the price per ounce of each bag? (Round to the nearest cent.)

3-oz bag: _____

14-oz bag: _____

b. Which bag of chips is the better buy?

_____

47

323

LESSON 12·6   Looking Back on the World Tour

Math Message

It is time to complete the World Tour.

1. Fly to Washington, D.C., and then travel to your hometown. Mark the final leg of the tour on the Route Map on *Math Journal 2,* pages 330 and 331.

2. What is the total distance you have traveled?   _____ miles

3. The airline has given you a coupon for every 5,000 miles you have traveled. Suppose you did all your traveling by plane on the same airline. How many coupons have you earned on the World Tour?   _____ coupons

4. You can trade in 5 coupons for one free round-trip ticket to fly anywhere in the continental United States. How many round-trip tickets have you earned on the World Tour?   _____ round-trip tickets

Refer to "My Country Notes" in your journals (*Math Journal 1,* pages 174–181 and *Math Journal 2,* pages 332–341) as you answer the following questions.

5. If you could travel all over the world for a whole year, what information would you need in order to plan your trip?

_____

_____

_____

_____

_____

_____

_____

_____

_____

LESSON 12·6 Looking Back on the World Tour *continued*

6. To which country would you most like to travel in your lifetime? Explain your answer.

_____

_____

_____

_____

7. On your travels, you would have the opportunity to learn about many different cultures. What would you want to share with people from other countries about *your* culture?

_____

_____

_____

_____

8. What are some things you have enjoyed on the World Tour?

_____

_____

_____

_____

9. What is something about the World Tour you would like to add or change?

_____

_____

_____

_____

LESSON 12·6 Rates

1. David Fischer of Chicago, Illinois, jumped rope 56 times in 60 seconds. If he could continue at this rate, about how many jumps would he do in 13 minutes?

   About _____ jumps

   Source: Guinness World Records, www.guinnessworldrecords.com

2. Some scientists recommend that the average person drink about 8 eight-ounce cups of water per day. At this rate, about how many cups of water would a person drink in 2 weeks?

   About _____ cups

3. On average, Americans eat about 20 pounds of pasta per year.

   a. Does this 20 pounds per year seem reasonable to you? _____

   Explain. _____

   _____

   b. At this rate, about how many pounds of pasta would an American eat in 23 years?

   About _____ pounds

   Source: Sicilian Culture, www.sicilianculture.com/food/pasta.htm

4. Thirty-six buses would be needed to carry the passengers and crew of three 747 jumbo jets.

   a. Fill in the rate table.

Buses			36	180	252
Jets	1	2	3		

   b. How many jets would be needed to carry the passengers on 264 buses? _____

5. A man in India grew one of his thumbnails until it was 114 centimeters long. Fingernails grow about 2.5 centimeters each year. At this rate, about how many years did it take the man in India to grow his thumbnail?

   About _____ years

LESSON 12·6

Math Boxes

1. a. Complete the table.

Number of Days	365			
Number of Years	1	2	9	12

    b. How many days are in 92 years? _____ days

SRB 47

2. Give other names for each measure.

    a. 1 gal

    _____ _____ _____

    b. 1 qt

    _____ _____ _____

    c. 500 mL

    _____ _____

SRB 137

3. Find the solution of each open sentence.

    a. $y + (-8) = -23$ $y =$ _____

    b. $12 + j = -5$ $j =$ _____

    c. $35 + r = 25$ $r =$ _____

    d. $c + (-115) = -144$ $c =$ _____

SRB 148

4. Complete the name-collection box.

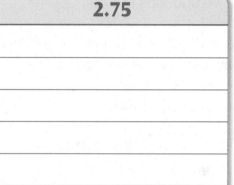

2.75

SRB 149

5. How likely is it that Raj will pick a 3 of clubs from a regular deck of playing cards without looking? Fill in the circle next to the best answer.

   Ⓐ   impossible

   Ⓑ   very unlikely

   Ⓒ   unlikely

   Ⓓ   very likely

SRB 80

6. Calculate.

    a. 10% of 860 = _____

    b. 5% of 220 = _____

    c. 75% of _____ = 12

    d. _____% of 87 = 43.5

    e. _____% of 60 = 18

    f. _____% of 145 = 36.25

SRB 38 39

LESSON 12·7   Math Boxes

---

1. Multiply. Show your work.

79 * 405 = _____

---

2. a. Complete the table.

Number of Inches			144	192	324
Number of Feet	1	2	9	12	

b. How many inches are in 11 feet?

_____ inches

---

3. It is estimated that in an average lifetime of 75 years, a person takes about 50,000 trips in a car.

a. At that rate, about how many times a day would a person ride in a car?

_____ times

b. Does this number make sense to you?

_____

---

4. Subtract.

a. −$75 − (+$25) = _____

b. −$45 − (−$30) = _____

c. −$60 − (+$60) = _____

d. $55 − (−$25) = _____

e. $300 − (−$100) = _____

---

5. Calculate the volume.

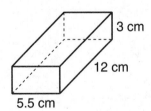

3 cm

12 cm

5.5 cm

Number model: _____

Volume = _____ cm³

---

6. A 10-ounce can of peas costs $0.55. A 16-ounce can of peas costs $1.19.

a. What is the price per ounce for each can? (Round to the nearest cent.)

10-oz can: _____

16-oz can: _____

b. Which can of peas is the better buy?

_____

My Route Log

Date	Country	Capital	Air distance from last capital	Total distance traveled so far
	1 U.S.A.	Washington, D.C.		
	2 Egypt	Cairo		
	3			
	4			
	5			
	6			
	7			
	8			
	9			
	10			
	11			
	12			
	13			
	14			
	15			
	16			
	17			
	18			
	19			
	20			

Route Map

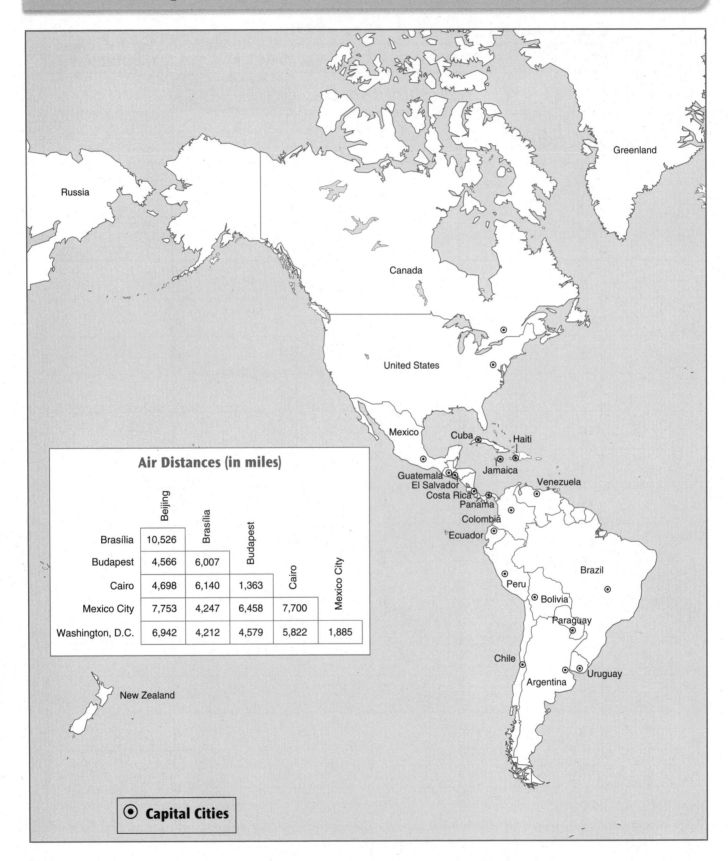

Air Distances (in miles)

	Beijing	Brasília	Budapest	Cairo	Mexico City
Brasília	10,526				
Budapest	4,566	6,007			
Cairo	4,698	6,140	1,363		
Mexico City	7,753	4,247	6,458	7,700	
Washington, D.C.	6,942	4,212	4,579	5,822	1,885

Russia

Greenland

Canada

United States

Mexico
Cuba
Haiti
Guatemala
Jamaica
El Salvador
Costa Rica
Venezuela
Panama
Colombia
Ecuador
Brazil
Peru
Bolivia
Paraguay
Chile
Uruguay
Argentina

New Zealand

⊙ **Capital Cities**

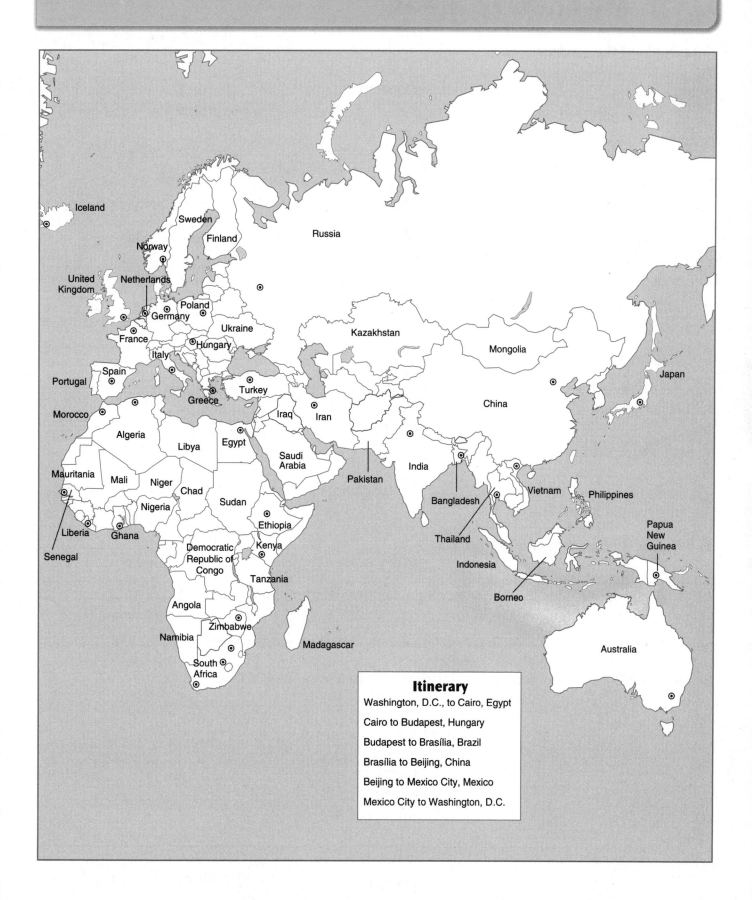

Itinerary

Washington, D.C., to Cairo, Egypt

Cairo to Budapest, Hungary

Budapest to Brasília, Brazil

Brasília to Beijing, China

Beijing to Mexico City, Mexico

Mexico City to Washington, D.C.

LESSON 7·2   My Country Notes

A. Facts about the country

_____ is located in _____.
name of country   name of continent

1. It is bordered by _____
countries, bodies of water

_____.

2. Population: _____ Area: _____ square miles

3. Languages spoken: _____

_____

4. Monetary unit: _____

5. Exchange rate (optional): 1 _____ = _____

B. Facts about the capital of the country

_____ Population: _____
name of capital

1. When it is noon in my hometown, it is _____ in _____.
time (A.M. or P.M.?)   name of capital

2. In _____/_____, the average high temperature in _____
month   month   name of capital

is about _____°F. The average low temperature is about _____°F.

3. What kinds of clothes should I pack for my visit to this capital? Why?

_____

_____

_____

_____

LESSON 7·2 My Country Notes *continued*

4. Turn to the Route Map found on journal pages 330 and 331.
   Draw a line from the last city you visited to the capital of this country.

5. If your class is using the Route Log, record the information on journal page 329 or *Math Masters,* page 421.

6. Can you find any facts on pages 302–305 in your *Student Reference Book* that apply to this country? For example, is one of the 10 tallest mountains in the world located in this country? List all the facts you can find.

_____

_____

_____

_____

_____

C. My impressions about the country

Do you know anyone who has visited or lived in this country? If so, ask that person for an interview. Read about the country's customs and about interesting places to visit there. Use encyclopedias, travel books, the travel section of a newspaper, or library books. Try to get brochures from a travel agent. Then describe below some interesting things you have learned about this country.

_____

_____

_____

_____

_____

_____

_____

My Country Notes

A. Facts about the country

_____ is located in _____ .
              name of country   name of continent

1. It is bordered by _____
                                             countries, bodies of water

_____ .

2. Population: _____   Area: _____ square miles

3. Languages spoken: _____

_____

4. Monetary unit: _____

5. Exchange rate (optional): 1 _____ = _____

B. Facts about the capital of the country

_____   Population: _____
          name of capital

1. When it is noon in my hometown, it is _____ in _____ .
                                       time (A.M. or P.M.?)   name of capital

2. In _____/_____ , the average high temperature in _____
       month   month   name of capital

is about _____°F. The average low temperature is about _____°F.

3. What kinds of clothes should I pack for my visit to this capital? Why?

_____

_____

_____

_____

LESSON 8·4   My Country Notes *continued*

4. Turn to the Route Map found on journal pages 330 and 331.
   Draw a line from the last city you visited to the capital of this country.

5. If your class is using the Route Log, record the information on journal page 329 or *Math Masters,* page 421.

6. Can you find any facts on pages 302–305 in your *Student Reference Book* that apply to this country? For example, is one of the 10 tallest mountains in the world located in this country? List all the facts you can find.

_____

_____

_____

_____

c. My impressions about the country

Do you know anyone who has visited or lived in this country? If so, ask that person for an interview. Read about the country's customs and about interesting places to visit there. Use encyclopedias, travel books, the travel section of a newspaper, or library books. Try to get brochures from a travel agent. Then describe below some interesting things you have learned about this country.

_____

_____

_____

_____

_____

_____

LESSON 9·5   My Country Notes

A. Facts about the country

_____ is located in _____.
name of country   name of continent

1. It is bordered by _____
countries, bodies of water

_____.

2. Population: _____   Area: _____ square miles

3. Languages spoken: _____

_____

4. Monetary unit: _____

5. Exchange rate (optional): 1 _____ = _____

B. Facts about the capital of the country

_____   Population: _____
name of capital

1. When it is noon in my hometown, it is _____ in _____.
time (A.M. or P.M.?)   name of capital

2. In _____/_____, the average high temperature in _____
month   month   name of capital

is about _____°F. The average low temperature is about _____°F.

3. What kinds of clothes should I pack for my visit to this capital? Why?

_____

_____

_____

_____

LESSON 9·5   My Country Notes *continued*

4. Turn to the Route Map found on journal pages 330 and 331. Draw a line from the last city you visited to the capital of this country.

5. If your class is using the Route Log, record the information on journal page 329 or *Math Masters,* page 421.

6. Can you find any facts on pages 302–305 in your *Student Reference Book* that apply to this country? For example, is one of the 10 tallest mountains in the world located in this country? List all the facts you can find.

_____

_____

_____

_____

_____

C. My impressions about the country

Do you know anyone who has visited or lived in this country? If so, ask that person for an interview. Read about the country's customs and about interesting places to visit there. Use encyclopedias, travel books, the travel section of a newspaper, or library books. Try to get brochures from a travel agent. Then describe below some interesting things you have learned about this country.

_____

_____

_____

_____

_____

_____

_____

LESSON 10·3 My Country Notes

A. Facts about the country

_____ is located in _____.

name of country   name of continent

1. It is bordered by _____

countries, bodies of water

_____.

2. Population: _____ Area: _____ square miles

3. Languages spoken: _____

_____

4. Monetary unit: _____

5. Exchange rate (optional): 1 _____ = _____

B. Facts about the capital of the country

_____ Population: _____

name of capital

1. When it is noon in my hometown, it is _____ in _____.

time (A.M. or P.M.?)   name of capital

2. In _____/_____, the average high temperature in _____

month   month   name of capital

is about _____°F. The average low temperature is about _____°F.

3. What kinds of clothes should I pack for my visit to this capital? Why?

_____

_____

_____

_____

LESSON 10·3 **My Country Notes** *continued*

4. Turn to the Route Map found on journal pages 330 and 331.
Draw a line from the last city you visited to the capital of this country.

5. If your class is using the Route Log, record the information on journal page 329 or *Math Masters,* page 421.

6. Can you find any facts on pages 302–305 in your *Student Reference Book* that apply to this country? For example, is one of the 10 tallest mountains in the world located in this country? List all the facts you can find.

_____

_____

_____

_____

_____

c. My impressions about the country

Do you know anyone who has visited or lived in this country? If so, ask that person for an interview. Read about the country's customs and about interesting places to visit there. Use encyclopedias, travel books, the travel section of a newspaper, or library books. Try to get brochures from a travel agent. Then describe below some interesting things you have learned about this country.

_____

_____

_____

_____

_____

_____

_____

LESSON 11·1 # My Country Notes

A. Facts about the country

_____ is located in _____.
name of country   name of continent

1. It is bordered by _____
countries, bodies of water

_____.

2. Population: _____   Area: _____ square miles

3. Languages spoken: _____

_____

4. Monetary unit: _____

5. Exchange rate (optional): 1 _____ = _____

B. Facts about the capital of the country

_____   Population: _____
name of capital

1. When it is noon in my hometown, it is _____ in _____.
time (A.M. or P.M.?)   name of capital

2. In _____/_____, the average high temperature in _____
month   month   name of capital

is about _____°F. The average low temperature is about _____°F.

3. What kinds of clothes should I pack for my visit to this capital? Why?

_____

_____

_____

_____

LESSON
11·1   **My Country Notes** *continued*   341

4. Turn to the Route Map found on journal pages 330 and 331.
    Draw a line from the last city you visited to the capital of this country.

5. If your class is using the Route Log, record the information on journal page 329 or
    Math Masters, page 421.

6. Can you find any facts on pages 302–305 in your *Student Reference Book* that
    apply to this country? For example, is one of the 10 tallest mountains in the world
    located in this country? List all the facts you can find.

    _____

    _____

    _____

    _____

    _____

C. My impressions about the country

Do you know anyone who has visited or lived in this country? If so, ask that person
for an interview. Read about the country's customs and about interesting places to
visit there. Use encyclopedias, travel books, the travel section of a newspaper, or
library books. Try to get brochures from a travel agent. Then describe below some
interesting things you have learned about this country.

    _____

    _____

    _____

    _____

    _____

    _____

    _____

Equivalent Names for Fractions

Fraction	Equivalent Fractions	Decimal	Percent
$\frac{0}{2}$		0	0%
$\frac{1}{2}$	$\frac{2}{4}, \frac{3}{6}$		
$\frac{2}{2}$		1	100%
$\frac{1}{3}$			
$\frac{2}{3}$			
$\frac{1}{4}$			
$\frac{3}{4}$			
$\frac{1}{5}$			
$\frac{2}{5}$			
$\frac{3}{5}$			
$\frac{4}{5}$			
$\frac{1}{6}$			
$\frac{5}{6}$			
$\frac{1}{8}$			
$\frac{3}{8}$			
$\frac{5}{8}$			
$\frac{7}{8}$			

Equivalent Names for Fractions *continued*

Fraction	Equivalent Fractions	Decimal	Percent
$\frac{1}{9}$			
$\frac{2}{9}$			
$\frac{4}{9}$			
$\frac{5}{9}$			
$\frac{7}{9}$			
$\frac{8}{9}$			
$\frac{1}{10}$			
$\frac{3}{10}$			
$\frac{7}{10}$			
$\frac{9}{10}$			
$\frac{1}{12}$			
$\frac{5}{12}$			
$\frac{7}{12}$			
$\frac{11}{12}$			

PROJECT 1 | # U.S. Traditional Addition 1

Algorithm Project 1

Use any strategy to solve the problem.

1. There are 279 boys and 347 girls at a school assembly. How many students are at the assembly?

   _____ students

Use U.S. traditional addition to solve each problem.

2.   559
     + 72
    —————

3.   3,743
     + 5,106
    —————

4.   328
     + 474
    —————

5. $1,885 + 6,167 =$ _____

6. _____ $= 456 + 198 + 618$

7. $5,506 + 4,677 =$ _____

U.S. Traditional Addition 2

Algorithm Project 1

Use U.S. traditional addition to solve each problem.

1. From Monday through Friday, Peng read
   388 pages of a book. On Saturday and
   Sunday, he read 159 more pages. How many
   pages did Peng read during the week?

   _____ pages

---

2.   633
         92
   +   48

3.   905
     +   496

---

4.   2,553
   +   6,424

5. 5,714 + 5,789 = _____

---

6. _____ = 4,343 + 526

7. 3,766 + 9,469 = _____

Algorithm Project 1

Use U.S. traditional addition to solve each problem.

1. Hiroshi had $356 in his bank account this morning. This afternoon he deposited $85 into the account. How much is in Hiroshi's account now?

$_____

2. Write a number story for 448 + 375.
Solve your number story.

_____

_____

_____

_____

Fill in the missing digits in the addition problems.

3.
```
    □ □ □
    5 6 9 6
  + 3 6 7 8
  ─────────
  □   3 □ 4
```

4.
```
          □
    6 3 5 2
  + □ 4 9 □
  ─────────
    7 8 □ 5
```

5.
```
    □ □
    4 9 4
  + 6 2 7
  ───────
  1 □ 2 □
```

6.
```
    1 □ 1
    9 9 8 □
  +   1 4 9
  ─────────
  1 □ 1 □ 5
```

PROJECT 1 — U.S. Traditional Addition 4

Algorithm Project 1

Use U.S. traditional addition to solve each problem.

1. Sara and James ran for school president.
   In the election, 529 students voted for Sara,
   and 378 voted for James. How many students
   voted in the election?

   _____ students

2. Write a number story for 483 + 577.
   Solve your number story.

   _____

   _____

   _____

   _____

Fill in the missing digits in the addition problems.

3.
```
      1   1
      5   6   3
  +   2   9   ☐
    ☐   ☐   2
```

4.
```
    1   1   1
    8   9   ☐   9
  +     ☐   0   2
  ☐     1   0   1
```

5.
```
    ☐   ☐   ☐
    2   8   5   8
  + 7   4   4   7
  1   ☐   3   ☐   ☐
```

6.
```
          ☐   1
    4   0   0   4
  + 8   6   9   ☐
  1   ☐   ☐   0   0
```

PROJECT 2

U.S. Traditional Addition: Decimals 1

Algorithm Project 2

Use any strategy to solve the problem.

1. Angela spent $2.62 at the craft store. She spent $3.94 at the fabric store. How much money did Angela spend in all?

   $ _____

Use U.S. traditional addition to solve each problem.

2. $7.69 + 38.5 = $ _____

3. _____ $= 6.48 + 29.6$

4. $\$9.59 + \$0.45 = \$$ _____

5. $\$30.45 + \$65.99 = \$$ _____

6. $54.11 + 9.2 = $ _____

7. _____ $= 2.88 + 83.09$

PROJECT 2 | # U.S. Traditional Addition: Decimals 2

Algorithm Project 2

Use U.S. traditional addition to solve each problem.

1. José had $5.98 in his wallet. He found 75¢ under his bed. How much money does José have now?

   $ _____

---

2. 3.9 + 4.48 = _____

3. 0.8 + 9.94 = _____

---

4. _____ = 6.76 + 28.18

5. 1.09 + 24.58 = _____

---

6. _____ = 1.03 + 52.81

7. 3.8 + 77.92 = _____

PROJECT 2   ## U.S. Traditional Addition: Decimals 3

Algorithm Project 2

Use U.S. traditional addition to solve each problem.

1. There is a flower growing in Kayla's garden. It was
    22.48 centimeters tall. In three months, it grew
    8.6 centimeters. How tall is the flower now?

    _____ centimeters

2. Write a number story for $3.80 + $5.12.
    Solve your number story.

    _____

    _____

    _____

    _____

Fill in the missing digits in the addition problems.

3.
```
        1   1
        3 . 8   5
    +     6 .[ ] 7
    ─────────────
    [ ][ ] . 5 [ ]
```

4.
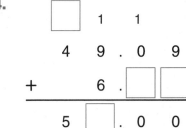
```
    [ ]   1   1
      4 9 . 0   9
    +     6 .[ ][ ]
    ─────────────
      5 [ ] . 0   0
```

5.
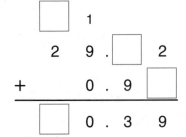
```
    [ ]     1
      2 9 .[ ] 2
    +     0 . 9 [ ]
    ─────────────
    [ ]   0 . 3   9
```

6.
```
          1   1
      7 [ ] . 4 [ ]
    + 1 2 . 8   6
    ─────────────
    [ ]   4 .[ ] 0
```

PROJECT 2

U.S. Traditional Addition: Decimals 4

Algorithm Project 2

Use U.S. traditional addition to solve each problem.

1. Surina and Lee are saving their money. Surina has $18.63. Lee has $24.81. How much money do they have altogether?

   $_____

2. Write a number story for 9.8 + 48.36.
   Solve your number story.

   _____

   _____

   _____

   _____

Fill in the missing digits in the addition problems.

3.
```
    ☐ ☐
  5 0 . 3  5
+     9 . 7  0
  ─────────────
  ☐ 0 . 0  ☐
```

4.
```
          1
      9 . 1   8
+     2 . ☐   ☐
  ─────────────
 ☐ ☐ . 9  1
```

5.
```
   1        1
   7  9 . 0  7
+  4  4 . ☐  5
 ──────────────
 ☐  2 ☐ . 4  ☐
```

6.
```
  1  1    ☐
  2  5 . 3  2
+ 2  ☐ . 7  9
 ──────────────
 ☐  0 . 1  ☐
```

Go to www.everydaymathonline.com for additional practice pages.

U.S. Traditional Subtraction 1

Algorithm Project 3

Use any strategy to solve the problem.

1. A store has 625 shirts and 379 pairs of pants. How many more shirts does the store have?

   _____ shirts

Use U.S. traditional subtraction to solve each problem.

2.   325
    − 68

3.   613
    − 249

4.   1,544
    − 749

5. 3,651 − 1,995 = _____

6. _____ = 506 − 187

7. 7,003 − 4,885 = _____

PROJECT 3

U.S. Traditional Subtraction 2

Algorithm Project 3

Use U.S. traditional subtraction to solve each problem.

1. The drive to Yuri's grandmother's house is 642 miles. Yuri's family has driven 484 miles so far. How many miles do they have left to drive?

   _____ miles

2.
```
   860
 −  86
```

3.
```
   707
 − 389
```

4.
```
   595
 − 397
```

5. _____ = 6,113 − 876

---

6. _____ = 4,552 − 1,688

7. 8,207 − 3,579 = _____

PROJECT 3 — **U.S. Traditional Subtraction 3**

Algorithm Project 3

Use U.S. traditional subtraction to solve each problem.

1. Althea has 233 bean-bag animals.
   79 of them are bears. How many of her
   bean-bag animals are not bears?

   _____ bean-bag animals

2. Write a number story for 505 − 267.
   Solve your number story.

   _____

   _____

   _____

   _____

Fill in the missing numbers in the subtraction problems.

3.
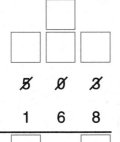

4.

```
   7  □  □  11
   8̸  0̸  4̸  1̸
 − 3  6  5  3
 ─────────────
   □  □  8  □
```

5.
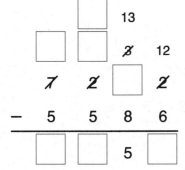

6.

```
      □
   7  □  □
   8̸  3̸  4̸
 − 5  □  5
 ───────────
   □  6  9
```

PROJECT 3

U.S. Traditional Subtraction 4

Algorithm Project 3

Use U.S. traditional subtraction to solve each problem.

1. Shane has $278 in his bank account. Caitlin has $425 in her bank account. How much more does Caitlin have in her account?

   $ _____

2. Write a number story for 503 − 347.
   Solve your number story.

   _____

   _____

   _____

   _____

Fill in the missing numbers in the subtraction problems.

3.
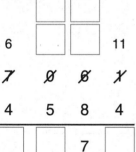

4.
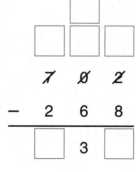

5.
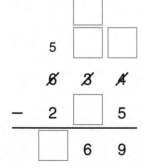

6.

Go to www.everydaymathonline.com for additional practice pages.

PROJECT 4

U.S. Traditional Subtraction: Decimals 1

Algorithm Project 4

Use any strategy to solve the problem.

1. Seth paid $6.72 for his lunch. Lily paid $3.79 for her lunch. How much more did Seth's lunch cost?

   $_____

Use U.S. traditional subtraction to solve each problem.

2. $9.75 - 4.32 =$ _____

3. $5.06 - 2.49 =$ _____

4. _____ $= 8.2 - 5.36$

5. $\$34.27 - \$16.38 = \$$_____

6. _____ $= 50.08 - 27.39$

7. $6.35 - 2.37 =$ _____

PROJECT 4

U.S. Traditional Subtraction: Decimals 2

Algorithm Project 4

Use U.S. traditional subtraction to solve each problem.

1. Joanna had $73.48 in her bank account.
   She wrote a check for $25.69. How much
   money is in her bank account now?

   $ _____

---

2. $6.04 - 2.75 =$ _____

3. $8.73 - 4.21 =$ _____

---

4. _____ $= 5.63 - 2.64$

5. $31.5 - 7.82 =$ _____

---

6. $ _____ $= \$45.26 - \$26.37$

7. $60.08 - 43.29 =$ _____

PROJECT 4

U.S. Traditional Subtraction: Decimals 3

Algorithm Project 4

Use U.S. traditional subtraction to solve each problem.

1. Riley bought two card games at the store. The total
   cost (before tax) was $9.25. One game cost $3.89.
   How much did the other game cost?

   $_____

2. Write a number story for $38.42 − $19.76.
   Solve your number story.

   _____

   _____

   _____

   _____

Fill in the missing numbers in the subtraction problems.

3.

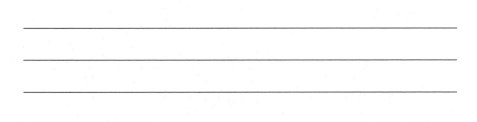

```
      □
    □ □ □
    7 . 0̸ 0̸
  − 3 . 8  9
  ─────────
    □ . 1 □
```

4.

```
        □
    4  □  11
    5̸ . 0̸  1̸
  − □ . 6  □
  ─────────
    1 . □  3
```

5.

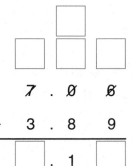

```
      □ □
   □  1̸0  □  □
   6̸  0̸ . 0̸  7
 − □  □ .  2  8
  ──────────────
   1  6 . □  9
```

6.

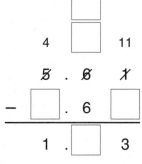

```
     □  □
   3  4̸  □  10
   4̸  5̸ . 4̸  □
 − □  8 .  8  5
  ──────────────
   2  □ . □  5
```

PROJECT 4

U.S. Traditional Subtraction: Decimals 4

Algorithm Project 4

Use U.S. traditional subtraction to solve each problem.

1. Quinn has two pieces of ribbon. The yellow ribbon is 12.42 meters long. The pink ribbon is 16.75 meters long. How much shorter is the yellow ribbon?

_____ meters

2. Write a number story for 7.63 − 1.84.
   Solve your number story.

   _____

   _____

   _____

   _____

Fill in the missing numbers in the subtraction problems.

3.
```
        □ □
   2  5̶ □  10
   3̶  6̶ . 4̶  □
 −    □  7 . 9  5
   _____
      □ . □     5
```

4.
```
         □   □
    □  1̶0  □   □
    4̶  0̶ . 0̶   6̶
 −  □  □ .  1   7
   _____
       6 . □    9
```

5.
```
        □
    7  □  11
    8̶ . 3̶  1̶
 −  □ . 3  □
   _____
    2 . □  4
```

6.
```
          □
     □  □  □
     9̶ . 0̶  4̶
 −   2 . 7  7
   _____
     □ . 2  □
```

U.S. Traditional Multiplication 1

Algorithm Project 5

Use any strategy to solve the problem.

1. Mountain View Elementary School held a food drive. Each student donated 4 cans of food. There are 676 students at the school. How many cans of food did the students donate altogether?

_____ cans

Use U.S. traditional multiplication to solve each problem.

2. 2 * 413 = _____

3. 265 * 4 = _____

4. _____ = 46 * 307

5. 278 * 43 = _____

6. 18 * 72 = _____

7. _____ = 459 * 40

U.S. Traditional Multiplication 2

Algorithm Project 5

Use U.S. traditional multiplication to solve each problem.

1. The Riveras' cornfield has 75 rows. Each row contains 256 corn plants. How many corn plants do the Riveras have in all?

_____ corn plants

2. $64 * 6 =$ _____

3. $213 * 30 =$ _____

4. $492 * 8 =$ _____

5. $70 * 572 =$ _____

6. $3 * 359 =$ _____

7. _____ $= 63 * 36$

PROJECT 5

U.S. Traditional Multiplication 3

Algorithm Project 5

Use U.S. traditional multiplication to solve each problem.

1. A machine can fill 258 bottles of juice per
   minute. How many bottles can the machine
   fill in 45 minutes?

   _____ bottles

2. Write a number story for 725 * 6.
   Solve your number story.

   _____

   _____

   _____

   _____

Fill in the missing digits in the multiplication problems.

3.
```
       [ ] 5
     4   2   9
  *          6
  2 [ ] 7 [ ]
```

4.
```
        3   2
        4   3
        3   6   5
   *        5   7
       2   5 [ ] 5
 + 1 [ ] [ ] 5   0
     2   0 [ ] 0   5
```

5.
```
            [ ]
          2
        6   4
   *      4 [ ]
      3   8   4
 + [ ] [ ] 6   0
   [ ] 9   4 [ ]
```

PROJECT 5 **U.S. Traditional Multiplication 4**

Algorithm Project 5

Use U.S. traditional multiplication to solve each problem.

1. The zebra at the city zoo weighs 627 pounds. The hippopotamus weighs 5 times as much as the zebra. How much does the hippopotamus weigh?

   _____ pounds

2. Write a number story for 584 * 23. Solve your number story.

   _____

   _____

   _____

   _____

Fill in the missing digits in the multiplication problems.

3.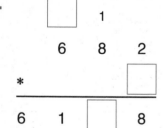

4.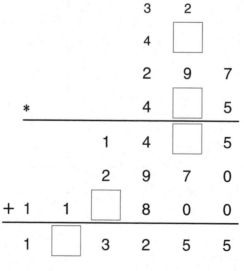

5.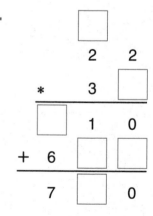

PROJECT 6

U.S. Traditional Multiplication: Decimals 1

Algorithm Project 6

Use any strategy to solve the problem.

1. A turkey sandwich at Jason's Deli costs $5.98.
   What is the cost of 4 turkey sandwiches?

   $ _____

Use U.S. traditional multiplication to solve each problem. Use estimation
or count decimal places to place the decimal point in your answers.

2. $12.64 * 5 =$ _____

3. $\$9.12 * 23 = \$$ _____

4. $\$$ _____ $= 86 * \$0.57$

5. $3 * \$45.80 = \$$ _____

6. _____ $= 50.7 * 65$

7. $426 * 5.3 =$ _____

PROJECT 6

U.S. Traditional Multiplication: Decimals 2

Algorithm Project 6

Use U.S. traditional multiplication to solve each problem. Use estimation
or count decimal places to place the decimal point in your answers.

1. Find the area of the rectangle.

_____ m²

5 m [   ]

24.36 m

---

2. 18 * 30.09 = _____

3. $24.05 * 6 = $_____

---

4. _____ = 34 * 0.67

5. $8.53 * 76 = $_____

---

6. _____ = 2.3 * 5,084

7. $5.21 * 4 = $_____

PROJECT 6 ## U.S. Traditional Multiplication: Decimals 3

Algorithm Project 6

Use U.S. traditional multiplication to solve each problem. Use estimation
or count decimal places to place the decimal point in your answers.

1. The average weight of a beagle puppy at birth
   is about 0.25 kg. At 6 months, a male beagle
   can weigh about 32 times as much. About how
   much can a 6-month-old male beagle weigh?

   _____ kg

2. Write a number story for 4.6 * 28.
   Solve your number story.

   _____

   _____

   _____

   _____

Fill in the missing digits in the multiplication problems.

3.

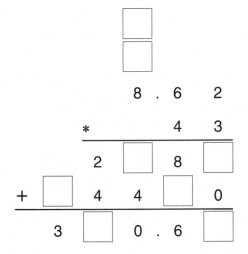

4.

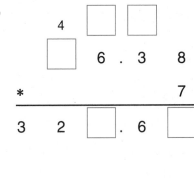

PROJECT 6 **U.S. Traditional Multiplication: Decimals 4**

Algorithm Project 6

Use U.S. traditional multiplication to solve each problem. Use estimation
or count decimal places to place the decimal point in your answers.

1. Alicia has 7 pieces of yarn. Each piece is
   3.65 meters long. What is the combined
   length of all 7 pieces?

   _____ m

2. Write a number story for 5 ∗ $48.30.
   Solve your number story.

   _____

   _____

   _____

   _____

Fill in the missing digits in the multiplication problems.

3.

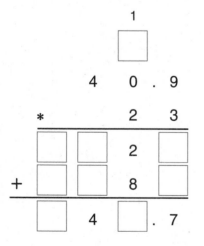

4.

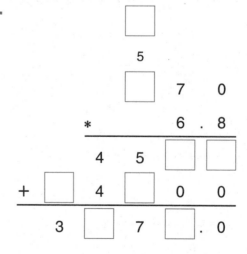

PROJECT 7 **Long Division with One-Digit Divisors**

Algorithm Project 7

Use any strategy to solve the problem.

1. The fourth-grade classes at Glendale School put on puppet shows for their families and friends. Ticket sales totaled $532, which the four classes are to share equally. How much should each class get?

   $_____

   Be ready to explain how you found your answer.

Use U.S. traditional long division to solve each problem.

2. 78 / 6 = _____

3. 288 / 8 = _____

4. _____ = 564 / 3

5. _____ = 763 / 7

PROJECT 7
Long Division with One-Digit Divisors *cont.*

Algorithm Project 7

6. 350 / 4 → _____

7. 802 / 9 → _____

8. _____ ← 869 / 7

9. _____ ← 874 / 5

Algorithm Project 7

10. Eight people visited a marine theme park. The total cost of the single-day admission tickets was $424. What was the cost per ticket?

$ _____

11. A national park charges an entrance fee of $3 per person. A school group visited the site. The cost was $288. How many people were in the school group?

_____ people

12. A family went on a six-day boat cruise. They sailed a total of 432 miles. They sailed the same distance each day. How far did they travel each day?

_____ miles

13. Four friends have birthdays in the same month. They decide to rent a hall to have a birthday party and split the cost evenly. The cost of renting the hall for one day is $172. How much did each friend pay?

$ _____

PROJECT 8

Long Division with Larger Dividends

Algorithm Project 8

Use any strategy to solve the problem.

1. Four friends were playing a board game. Jen had to leave to go to her piano lesson. The three other players decided to divide Jen's money equally. Jen had $4,353. How much should each of the three other players get?

   $ _____

   Be ready to explain how you got your answer.

_____

Use U.S. traditional long division to solve each problem.

2. $5,385 / 5 = $ _____

3. $7,896 / 6 = $ _____

4. _____ = 8,575 / 7

5. _____ = 8,127 / 3

Algorithm Project 8

Fill in the missing numbers.

6.

```
        1 □ 3 9
    5 ) 8 6 9 5
      - 5
        3 6
      - 3 5
          □ 9
        - 1 5
            4 □
          - 4 5
              0
```

7.

```
          5 □ □
    6 ) 3 2 5 2
      - □ 0
          2 □
        - 2 4
            1 2
          - 1 □
              0
```

8. Jai is saving money to go to sleep-away camp next summer. The total cost is $1,092. He is earning money by walking dogs in his neighborhood.

   a. At $4 per walk, how many dogs will Jai need to walk to earn $1,092?

      _____ dogs

   b. At $7 per walk, how many dogs will Jai need to walk to earn $1,092?

      _____ dogs

PROJECT 8

Long Division with Dollars and Cents

Algorithm Project 8

1. Dennis solved $9.45 / 7 like this.

   a. Study Dennis's work.

   b. Explain to your partner how he solved the problem.

```
     1.3 5
  7)9.4 5
   -7
    2 4
   -2 1
      3 5
     -3 5
        0
```

Solve these division problems using Dennis's method.

2. $8.92 / 4 = $_____

3. $7.56 / 6 = $_____

4. _____ = 15.76 / 8

5. _____ = 19.17 / 9

Fraction Cards 1

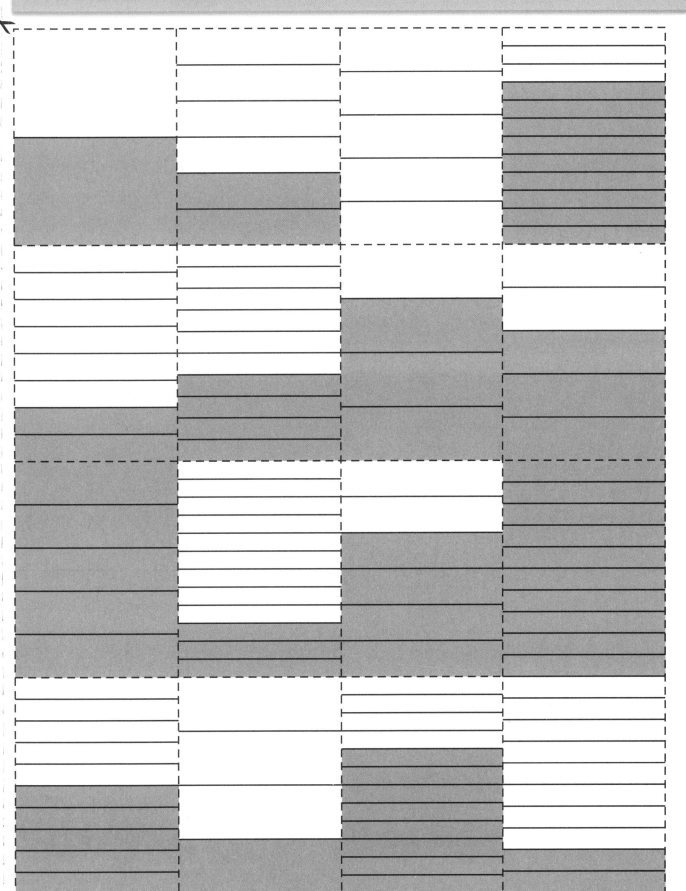

Fraction Cards 1

$$\frac{9}{} \qquad \frac{}{5} \qquad \frac{2}{} \qquad \frac{}{2}$$

$$\frac{3}{} \qquad \frac{}{4} \qquad \frac{4}{} \qquad \frac{2}{}$$

$$\frac{10}{} \qquad \frac{}{6} \qquad \frac{3}{} \qquad \frac{}{5}$$

$$\frac{2}{} \qquad \frac{}{12} \qquad \frac{1}{} \qquad \frac{}{10}$$

Activity Sheet 5

Fraction Cards 2

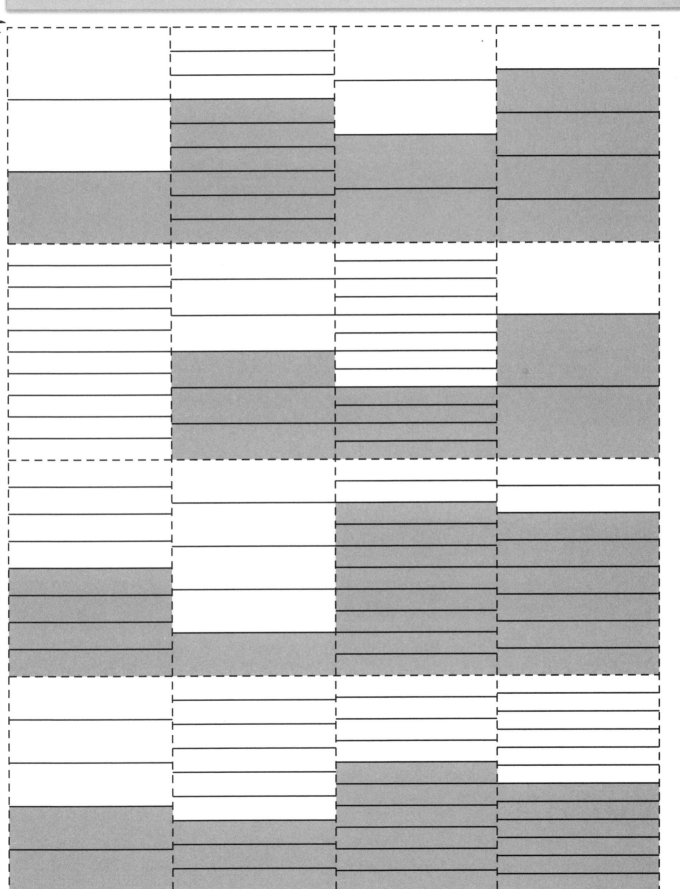

Fraction Cards 2

$$\frac{}{5} \qquad \frac{2}{} \qquad \frac{}{9} \qquad \frac{1}{}$$

$$\frac{}{3} \qquad \frac{4}{} \qquad \frac{3}{} \qquad \frac{0}{}$$

$$\frac{6}{} \qquad \frac{}{10} \qquad \frac{1}{} \qquad \frac{}{8}$$

$$\frac{}{12} \qquad \frac{6}{} \qquad \frac{}{9} \qquad \frac{2}{}$$

Activity Sheet 6

LESSON 10·2 Dart Game

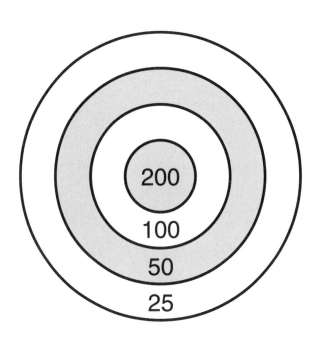

Scoreboard 1	
Player 1	Player 2

Scoreboard 2	
Player 1	Player 2

Scoreboard 3	
Player 1	Player 2

Activity Sheet 7

**LESSON
10·2**   *Pocket-Billiards Game*

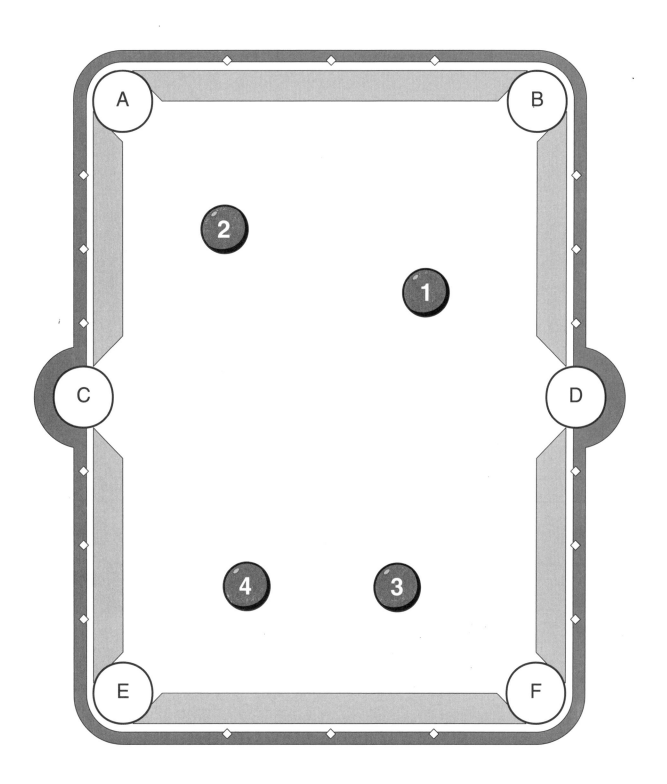